A Review of the Use of Science and Adaptive Management in California's Draft Bay Delta Conservation Plan

Panel to Review California's Draft Bay Delta Conservation Plan

Water Science and Technology Board

Ocean Studies Board

Division on Earth and Life Studies

NATIONAL RESEARCH COUNCIL
OF THE NATIONAL ACADEMIES

THE NATIONAL ACADEMIES PRESS
Washington, D.C.
www.nap.edu

THE NATIONAL ACADEMIES PRESS 500 Fifth Street, N.W. Washington, DC 20001

NOTICE: The project that is the subject of this report was approved by the Governing Board of the National Research Council, whose members are drawn from the councils of the National Academy of Sciences, the National Academy of Engineering, and the Institute of Medicine. The members of the panel responsible for the report were chosen for their special competences and with regard for appropriate balance.

Support for this study was provided by the Department of the Interior under contract no. 80221-A-G100. Any opinions, findings, conclusions, or recommendations expressed in this publication are those of the author(s) and do not necessarily reflect the views of the organizations or agencies that provided support for the project.

International Standard Book Number 13: 978-0-309-21231-1
International Standard Book Number 10:0-309-21231-6

Photo on the cover is courtesy of David Policansky.

Additional copies of this report are available from the National Academies Press, 500 5th Street, N.W., Lockbox 285, Washington, DC 20055; (800) 624-6242 or (202) 334-3313 (in the Washington metropolitan area); Internet, *http://www.nap.edu*.

THE NATIONAL ACADEMIES
Advisers to the Nation on Science, Engineering, and Medicine

The **National Academy of Sciences** is a private, nonprofit, self-perpetuating society of distinguished scholars engaged in scientific and engineering research, dedicated to the furtherance of science and technology and to their use for the general welfare. Upon the authority of the charter granted to it by the Congress in 1863, the Academy has a mandate that requires it to advise the federal government on scientific and technical matters. Dr. Ralph J. Cicerone is president of the National Academy of Sciences.

The **National Academy of Engineering** was established in 1964, under the charter of the National Academy of Sciences, as a parallel organization of outstanding engineers. It is autonomous in its administration and in the selection of its members, sharing with the National Academy of Sciences the responsibility for advising the federal government. The National Academy of Engineering also sponsors engineering programs aimed at meeting national needs, encourages education and research, and recognizes the superior achievements of engineers. Dr. Charles M. Vest is president of the National Academy of Engineering.

The **Institute of Medicine** was established in 1970 by the National Academy of Sciences to secure the services of eminent members of appropriate professions in the examination of policy matters pertaining to the health of the public. The Institute acts under the responsibility given to the National Academy of Sciences by its congressional charter to be an adviser to the federal government and, upon its own initiative, to identify issues of medical care, research, and education. Dr. Harvey V. Fineberg is president of the Institute of Medicine.

The **National Research Council** was organized by the National Academy of Sciences in 1916 to associate the broad community of science and technology with the Academy's purposes of furthering knowledge and advising the federal government. Functioning in accordance with general policies determined by the Academy, the Council has become the principal operating agency of both the National Academy of Sciences and the National Academy of Engineering in providing services to the government, the public, and the scientific and engineering communities. The Council is administered jointly by both Academies and the Institute of Medicine. Dr. Ralph J. Cicerone and Dr. Charles M. Vest are chair and vice chair, respectively, of the National Research Council.

www.national-academies.org

PANEL TO REVIEW CALIFORNIA'S
DRAFT BAY DELTA CONSERVATION PLAN *

HENRY J. VAUX, JR., Chair, Professor Emeritus, University of California
MICHAEL E. CAMPANA, Oregon State University, Corvallis
JEROME B. GILBERT, Consultant, Orinda, California
ALBERT E. GIORGI, BioAnalysts, Inc., Redmond, Washington
ROBERT J. HUGGETT, Professor Emeritus, College of William and Mary, Seaford, Virginia
CHRISTINE A. KLEIN, University of Florida College of Law, Gainesville, Florida
SAMUEL N. LUOMA, U.S. Geological Survey, Emeritus, Menlo Park, California
THOMAS MILLER, University of Maryland Center for Environmental Science, Chesapeake Biological Laboratory, Solomons
STEPHEN G. MONISMITH, Stanford University, California
JAYANTHA OBEYSEKERA, South Florida Water Management District, West Palm Beach
HANS W. PAERL, University of North Carolina, Chapel Hill
MAX J. PFEFFER, Cornell University, Ithaca, New York
DESIREE D. TULLOS, Oregon State University, Corvallis

NRC Staff

LAURA J. HELSABECK, Staff Officer
DAVID POLICANSKY, Scholar
STEPHEN D. PARKER, Director, Water Science and Technology Board
SUSAN ROBERTS, Director, Ocean Studies Board
ELLEN DE GUZMAN, Research Associate
SARAH BRENNAN, Senior Program Assistant

---

* Biographical information for panel members is in Appendix H. This project was organized and overseen by the NRC's Water Science and Technology Board (lead) and Ocean Studies Board, whose rosters are in Appendixes F and G, respectively.

Preface

This panel's review of the draft Bay Delta Conservation Plan (BDCP) has occurred alongside myriad activities in the Delta to facilitate a secure water future for California, including an environmental future, and alongside related activities of the National Research Council (NRC). I particularly want to make clear the distinction between the Delta Plan and the BDCP, and between this panel's report and two related NRC reports, one already published, one still in preparation.

The Delta Plan (formally the Delta Stewardship Plan) is a comprehensive umbrella plan mandated by the California Delta Protection Act of 2009 to advance the goals of improving the reliability of California's water supply and restoring, protecting, and enhancing the Delta ecosystem. It is overseen by the state of California and a broadly represented council of stakeholders as authorized by statute. Although the Delta Plan was not part of this review and is mentioned only incidentally in this report, it is related to the BDCP to some degree by intent and to some degree by statute (those relationships are briefly discussed in the body of this report). Readers should understand from the outset, however, that it is the BDCP, and only the BDCP that is reviewed in this report.

The related NRC activities are being conducted by the Committee on Sustainable Water and Environmental Management in the California Bay-Delta. The NRC appointed that committee in response to a request from Congress and the Department of the Interior to provide advice on two topics: (1) the scientific basis of actions identified in two biological opinions by the National Marine Fisheries Service and the U.S. Fish and Wildlife Service to protect threatened and endangered species in the Delta, and (2) how to most effectively incorporate science and adaptive management into a holistic program for managing and restoring the Delta. Advice on the first topic was provided in a report published in March 2010 titled *A Scientific Assessment of Alternatives for Reducing Water Management Effects on Threatened and Endangered Fishes in California's Bay-Delta*. The committee expects to release its advice on the second topic late in 2011.

While the committee was working on its second report, the U.S. Secretaries of Interior and Commerce asked the NRC to review the draft BDCP in terms of its use of science and adaptive management. In response, the NRC established a separate Panel to Review California's Draft Bay Delta Conservation Plan, which is the author of this report. Although there is considerable overlap between the membership of the committee and this panel, the two groups were appointed separately, have separate statements of task, and have worked independently of each other.

This report was reviewed in draft form by individuals chosen for their diverse perspectives and technical expertise in accordance with the procedures approved by the NRC's Report Review Committee. The purpose of this independent review is to provide candid and critical comments that will assist the NRC in making its published report as sound as possible, and to ensure that the report meets NRC institutional standards for objectivity, evidence, and responsiveness to the study charge. The review comments and draft manuscript remain confidential to protect the integrity of the deliberative process.

We thank the following for their review of this report: Frank Davis, University of California, Santa Barbara; Holly Doremus, University of California, Berkeley; Peter Gleick, Pacific Institute for Studies in Development, Environment, and Security; George Hornberger, Vanderbilt University; Cynthia Jones, Old Dominion University; Jay Lund, University of California, Davis; Judy Meyer, University of Georgia; and Lynn Scarlett, Resources for the Future.

Although these reviewers provided constructive comments and suggestions, they were not asked to endorse the report's conclusions and recommendations, nor did they see the final draft of the report before its release. The review of this report was overseen by Michael Kavanaugh, Geosyntec Consultants, who was appointed by the NRC's Report Review Committee and by Paul Risser, University of Oklahoma, who was appointed by the NRC's Division on Earth and Life Studies. They were responsible for ensuring that an independent examination of this report was conducted in accordance with NRC institutional procedures and that all review comments received full consideration. Responsibility for this report's final contents rests entirely with the authoring committee and the NRC.

Henry J. Vaux, Jr.
Chair
Panel to Review California's Draft Bay Delta Conservation Plan

Contents

Summary 1

1   INTRODUCTION 9

2   BACKGROUND 11

3   CRITICAL GAPS IN THE SCOPE OF THE DRAFT BDCP.............. 20
     The Lack of an Effects Analysis 21
     The Lack of Clarity as to the BDCP's Purpose........................... 25

4   USE OF SCIENCE IN THE BDCP... 29
     Incorporating Risk Analysis 30
     Integration of Climate Change Analysis..................................... 31
     A Framework for Linking Drivers and Effects 34
     Significant Environmental Factors Affecting Listed Species 35
     Synthesis.. 36
     The Relationship of the BDCP to Other Scientific Efforts 36

5   ADAPTIVE MANAGEMENT IN THE BDCP 38

6   MANAGEMENT FRAGMENTATION
     AND A LACK OF COHERENCE 45

7   IN CONCLUSION 50

References... 52

Appendix A   Statement of Task.. 63
Appendix B   BDCP Steering Committee Members and
              Planning Agreement Signature Dates..................... 64
Appendix C   BDCP Proposed Covered Species and
              Associated Habitats 64
Appendix D   Possible Causal Connections in Suppression of
              Populations of Endangered Suckers in
              Upper Klamath Lake 72
Appendix E   BDCP Adaptive Management Process
              Framework... 73
Appendix F   Water Science and Technology Board 74

Appendix G   Ocean Studies Board 75
Appendix H   Panel Biographical Information.. 76

Summary

The San Francisco Bay Delta Estuary (Delta, for short) is a large, complex estuarine ecosystem in California (Figure 1). It has been substantially altered by dikes, levees, channelization, pumps, human development, introduced species, dams on its tributary streams, and contaminants. The Delta supplies water from the state's wetter northern regions to the drier southern regions and also serves as habitat for many species, some of which are threatened and endangered. The restriction of water exports in an attempt to protect those species together with the effects of several dry years have exacerbated tensions over water allocation in recent years, and have led to various attempts to develop comprehensive plans to provide reliable water supplies and to protect the ecosystem.

One of those plans is the Bay Delta Conservation Plan (BDCP), the focus of this report. The BDCP is technically a habitat conservation plan (HCP), an activity provided for in the federal Endangered Species Act that protects the habitat of listed species in order to mitigate the adverse effects of a federal project or activity that incidentally "takes"[1] (includes actions that "harm" wildlife by impairing breeding, feeding, or sheltering behaviors) the listed species. It similarly is a natural community conservation plan (NCCP) under California's Natural Community Conservation Planning Act (NCCPA). It is intended to obtain long-term authorizations under both the state and federal endangered species statutes for proposed new water operations—primarily an "isolated conveyance structure," probably a tunnel, to take water from the northern part of the Delta for export to the south, thus reducing the need to convey water through the Delta and out of its southern end.

The U.S. Secretaries of the Interior and Commerce requested that the National Research Council (NRC) review the draft BDCP in terms of its use of science and adaptive-management (see Appendix A for the full statement of task). In response, the NRC established the Panel to Review California's Draft Bay Delta Conservation Plan, which prepared this report. The panel reviewed

---

[1] *Take* means "to harass, harm, pursue, hunt, shoot, wound, kill, trap, capture, or collect, or to attempt to engage in any such conduct." ESA, Section 3, 16 U.S.C. 1532.

Harm, within the statutory definition of "take" has been further defined by regulation: "Harm in the definition of take in the Act means an act which actually kills or injures wildlife. Such act may include significant habitat modification or degradation where it actually kills or injures wildlife by significantly impairing essential behavioral patterns, including breeding, feeding, or sheltering." 50 C.F.R. 17.3.

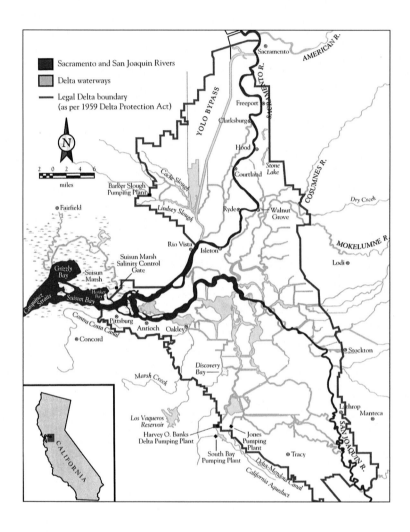

FIGURE 1. The Sacramento-San Joaquin Delta in California. San Francisco Bay, an integral part of the system, is just to the west. SOURCE: Reprinted, with permission, from Lund et al. (2010). Copyright by Public Policy Institute of California.

the draft BDCP, which was posted on the BDCP website: (*http://www.re-sources.ca.gov/bdcp/*) on November 18, 2010. [2] The panel determined that the draft BDCP is incomplete in a number of important areas and takes this opportunity to identify key scientific and structural gaps that, if addressed, could lead to a more successful and comprehensive final BDCP. Yet science alone cannot solve the Delta's problems. Water scarcity in California is very real, the situation is legally and politically complex, and many stakeholders have differing interests. The effective management of scarcity requires not only the best science and technology, but also consideration of public and private values, usually through political processes, to arrive at plans of action that are scientifically based but also incorporate and reflect the mix of differing personal and group values.

CRITICAL GAPS IN THE SCOPE OF THE DRAFT BDCP

At the outset of its review, the panel identified a problem with the geographical and hydrologic scope of the draft BDCP. The BDCP aims to address management and restoration of the San Francisco Bay Delta Estuary, an estuary that extends from the Central Valley to the mouth of San Francisco Bay. Thus, given that the BDCP describes a *bay* delta conservation plan, the omission of analyses of the effects of the BDCP efforts on San Francisco Bay (aside from Suisun Bay) is notable.

The Lack of an Effects Analysis

The draft BDCP describes an effects analysis as:

"the principal component of a habitat conservation plan. . . . The analysis includes the effects of the proposed project on covered species, including federally and state listed species, and other sensitive species potentially affected by the proposed project. The effects analysis is a systematic, scientific look at the potential impacts of a proposed project on those species and how those species would benefit from conservation actions." (draft BDCP, p. 5-2)

Clearly, such an effects analysis, which is in preparation, is intended to be the basis for the choice and details of those conservation actions. Its absence in the draft BDCP, therefore, is a critical gap in the science in the BDCP and the corresponding conservation actions. Nevertheless, the panel takes this opportunity

---

[2] BDCP (Bay Delta Conservation Plan Steering Committee). 2010. Bay Delta Conservation Plan Working Draft. November 18. Available online at: *http://www.resources.ca.gov/bdcp/*. Last accessed April 26, 2011.

to present its vision of a successful effects analysis, which includes an integrated description of the components of the system and how they relate to each other; a synthesis of the best available science; and a representation of the dynamic response of the system.

The term "effects analysis" also applies to an analysis of what is causing the listed (and other ecologically important) species to decline. In such a case, the logical sequence would be to perform the effects analysis on the causes of the species' declines, then design a proposed alternative to current operations to help reverse those declines, and then perform a second effects analysis on the probable effects of the proposed alternative. This aspect of an effects analysis is not mentioned in the current draft of the BDCP, and its absence brings the panel to a second critical gap in the scope of the draft BDCP, namely, a lack of clarity of the BDCP's purpose.

The Lack of Clarity as to the BDCP's Purpose

The legal framework underlying the BDCP is complex, as are the challenges of assembling such a large habitat conservation plan. Nonetheless, the BDCP's purpose or purposes need to be clearly stated, because their nature and interpretation are closely tied to the BDCP's scientific elements. The lack of clarity makes it difficult for this panel and the public to properly understand, interpret, and review the science that underlies the BDCP.

The central issue is to what extent the BDCP is only an application for a permit to incidentally take listed species, and to what extent it also is designed to achieve the two co-equal goals of providing for a more reliable water supply for the state of California and protecting, restoring, and enhancing the Delta ecosystem specified in recent California water legislation. To obtain an incidental take permit, it is logical to identify a proposed project or operation and design conservation methods to minimize and mitigate its adverse effects. But if the BDCP were largely a broader conservation program, designed to protect the ecosystem and provide a reliable water supply, then a more logical sequence would be to choose alternative projects or operating regimes only after the effects analysis was complete. Under that scenario, choosing the alternative first would be like putting the cart before the horse, or *post hoc* rationalization; in other words, choosing a solution before evaluating alternatives to reach a preferred outcome.

A related issue is the lack of consideration of alternatives to the preferred proposal (i.e., the isolated conveyance system). To the degree that the reasons for not considering alternatives have a scientific (as opposed to, for example, a financial) basis, their absence makes the BDCP's purpose less clear, and the panel's task more difficult.

THE USE OF SCIENCE AND SYNTHESIS IN THE BDCP

Many scientific efforts are and have been under way to understand and monitor hydrologic, geologic, and ecological interactions in the Delta, efforts that constitute the BDCP's scientific foundation. But overall it is not clear how the BDCP's authors synthesized the foundation material and systematically incorporated it into the decision-making process that led to the plan's conservation actions. For example, it is not clear how the Delta Regional Ecosystem Restoration Implementation Plan has been incorporated into the draft BDCP (see Appendix F of the draft BDCP). It also is not clear whether and how the draft BDCP incorporated the analyses for the Delta Risk Management Strategy and the framework developed by the Interagency Ecological Program related to factors affecting pelagic organism decline.

Furthermore, some of the scientific efforts related to the BDCP were incomplete at the time of this review. For example, warming, sea level rise, and changes in precipitation patterns and amounts will play a central role in Delta water allocation and its effects. Although the draft BDCP does mention incorporation of climate variability and change and model uncertainty, such information was not included in the draft BDCP that was provided.

Several other conservation efforts have been undertaken in the Delta in response to consultations with the National Marine Fisheries Service and the U.S. Fish and Wildlife Service concerning the potential for project operations (e.g., pumping) to jeopardize the listed species. The link between the BDCP and these other efforts is unclear. For example, the Delta Plan is a comprehensive conservation, restoration, and water-supply plan mandated in recent California legislation. That legislation also provided for potential linkage between the BDCP and the Delta Plan, but the draft BDCP does not make clear how this new relationship will be operationalized.

Much of the analysis of the factors affecting the decline of smelt and salmonids in the Delta has focused on water operations there, in particular, the pumping of water at the south end of the Delta for export to other regions. However, a variety of other significant environmental factors ("other stressors") have potentially large effects on the listed fishes. In addition, there remain considerable uncertainties surrounding the degree to which different aspects of flow management in the Delta, especially management of the salinity gradient, affect the survival of the listed fishes. Indeed, the significance and appropriate criteria for future environmental flow optimization have yet to be established, and are uncertain at best. The panel supports the concept of a quantitative evaluation of stressors, ideally using life-cycle models, as part of the BDCP.

The lack of clarity concerning the volume of water to be diverted is a major shortcoming of the BDCP. In addition, the BDCP provides little or no information about the reliability of supply for such a diversion or the different reliabilities associated with diversions of different volumes. It is nearly impossible to evaluate the BDCP without a clear specification of the volume(s) of water to be diverted, whose negative impacts the BDCP is intended to mitigate.

The draft BDCP is little more than a list of ecosystem restoration tactics and scientific efforts, with no clear over-arching strategy to tie them together or to implement them coherently to address mitigation of incidental take and achievement of the co-equal goals and ecosystem restoration. The relationships between scientific programs and efforts external to the BDCP and the BDCP itself are not clear. Furthermore scientific elements within the BDCP itself are not clearly related to each other. A systematic and comprehensive restoration plan needs a clearly stated strategic view of what each major scientific component of the plan is intended to accomplish and how this will be done. The separate scientific components should be linked, when relevant, and systematically incorporated into the BDCP. Also, a systematic and comprehensive plan should show how its (in this case, co-equal) goals are coordinated and integrated into a single resource plan and how this fits into and is coordinated with other conservation efforts in the Delta, for example, the broader Delta Plan.

ADAPTIVE MANAGEMENT

Numerous attempts have been made to develop and implement adaptive management strategies in environmental management, but many of them have not been successful, for a variety of reasons, including lack of resources; unwillingness of decision makers to admit to and embrace uncertainty; institutional, legal, and political preferences for known and predictable outcomes; the inherent uncertainty and variability of natural systems; the high cost of implementation; and the lack of clear mechanisms for incorporating scientific findings into decision making. Despite all of the above challenges, often there is no better option for implementing management regimes, and thus the panel concludes that the use of adaptive management is appropriate in the BDCP. However, the application of adaptive management to a large-scale problem like the one that exists in California's Bay-Delta will not be easy, quick, or inexpensive. The panel concludes that the BDCP needs to address these difficult problems and integrate conservation measures into the adaptive management strategy before there can be confidence in the adaptive management program. In addition, the above considerations emphasize the need for clear goals and integrated goals, which have not been provided by the draft BDCP. Although no adaptive management program can be fully described before it has begun, because such programs evolve as they are implemented, some aspects of the program could have been laid out more clearly than they have been.

Adaptive management requires a monitoring program to be in place. The draft BDCP does describe its plan for a monitoring program in considerable detail. However, given the lack of clarity of the BDCP's purpose and of any effects analysis, it is difficult to evaluate the motivation and purpose of the monitoring program. An effective monitoring program should be tied to the effects analysis, its purpose should be clear (e.g., to establish reference or baseline conditions, to detect trends, to serve as an early-warning system, to monitor management regimes for effectiveness), and it should include a mechanism for linking the information gained to operational decision making and to the monitoring itself. Those elements are not clearly described in the draft BDCP.

In 2009, the BDCP engaged a group of Independent Science Advisors to provide expertise on approaches to adaptive management. The panel concludes that the Independent Science Advisors provided a logical framework and guidance for the development and implementation of an appropriate adaptive management program for the BDCP. However, the draft BDCP lacks details to demonstrate that the adaptive management program is properly designed and follows the guidelines provided by the Independent Science Advisors. The panel further concludes that the BDCP developers could benefit significantly from adaptive management experiences in other large-scale ecosystem restoration efforts, such as the Comprehensive Everglades Restoration Program. The panel recognizes that no models exactly fit the Delta situation, but this should not prevent planners from using the best of watershed-restoration plans to develop an understandable, coherent, and data-based program to meet California's restoration and reliability goals. Even a soundly implemented adaptive management program is not a guarantee of achieving the BDCP's goals, however, because many factors outside the purview of the adaptive-management program may hinder restoration. However, a well-designed and implemented adaptive management program should make the BDCP's success more likely.

MANAGEMENT FRAGMENTATION AND A LACK OF COHERENCE

The absence of scientific synthesis in the draft BDCP draws attention to the fragmented system of management under which the plan was prepared—a management system that lacks coordination among entities and clear accountability. No one public agency, stakeholder group or individual has been made accountable for the coherence, thoroughness, and effectiveness of the final product. Rather, the plan appears to reflect the differing perspectives of federal, state, and local agencies, and the many stakeholder groups involved. Although this is not strictly a scientific issue, fragmented management is a significant impediment to the use and inclusion of coherent science in future iterations of the BDCP. Different science bears on the missions of the various public agencies, and different stakeholders put differing degrees of emphasis on specific pieces of science. Unless the management structure is made more coherent and unified, the final product may continue to suffer from a lack of integration in an attempt to satisfy all discrete interests and not, as a result, the larger public interests.

IN CONCLUSION

The panel finds the draft BDCP to be incomplete or unclear in a variety of ways and places. The plan is missing the type of structure usually associated with current planning methods in which the goals and objectives are specified, alternative measures for achieving the objectives are introduced and analyzed, and a course of action is identified based on analytical optimization of economic, social, and environmental factors. Yet the panel underscores the importance of a credible and a robust BDCP in addressing the various water management problems that beset the Delta. A stronger, more complete, and more scientifically credible BDCP that effectively integrates and utilizes science could indeed pave the way toward the next generation of solutions to California's chronic water problems.

1

Introduction

The San Francisco Bay Delta Estuary encompasses the deltas of the Sacramento and San Joaquin Rivers as well as the eastern margins of San Francisco Bay. Although the area has been extensively modified over the past 150 years, it remains biologically diverse while functioning as a central element in California's water supply system. The Delta system is subject to several forces of change, including seismicity, land subsidence, sea level rise, and changes in flow magnitudes as well as such societal changes as increased urbanization, population growth, growing water demands, and changing agricultural practices. These changes threaten the integrity of the Delta and its capacity to function both as an important link in the state's water supply system and as habitat for many species, some of which are threatened and endangered. In anticipation of the need to manage and respond to changes that have already and are likely to beset the Delta, a variety of planning activities have been undertaken. One such activity entails the development of a Bay Delta Conservation Plan (BDCP) by a consortium of federal, state, and local government agencies, environmental organizations, water supply entities, and other interested parties as a habitat conservation plan (see Appendix B). The BDCP covers 11 fish, 6 mammal, 12 bird, 2 reptile, 3 amphibian, 8 invertebrate, and 21 plant species (see Appendix C).

The present volume is the report of a panel appointed by the National Research Council at the request of the U.S. Secretaries of Interior and Commerce to review a working draft of the BDCP, dated November 18, 2010.[3] Specifically, the panel was charged with providing a short report assessing the adequacy of the use of science and adaptive management in the draft BDCP (see Appendix A). The panel met on December 8 and 10, 2010 in San Francisco, California. On the first day the panel heard presentations from the various authors and sponsors of the draft BDCP and commentary from interested stakeholders. The panel spent the remainder of the meeting time as well as the intervening weeks examining, evaluating, and analyzing the draft BDCP. In the course of this review, the panel delved into supporting documents such as the Delta Risk Management Strategy and other relevant documents. This report refers to and comments on those documents in the context of the BDCP; however, this report is not a review of those documents.

The use of science has been emphasized in recent legislation, and science is

---

[3] BDCP (Bay Delta Conservation Plan Steering Committee). 2010. Bay Delta Conservation Plan Working Draft. November 18. Available online at: *http://www.resources.ca.gov/bdcp/*. Last accessed April 26, 2011.

undoubtedly essential to the development of Delta plans generally. But science is only a starting point in the development of an integrated watershed-based plan, and it must be broadly applied. Moreover, science by itself cannot generate solutions to the myriad problems of the Delta that will satisfy the interests of all parties. Water scarcity in California is very real and science is not necessarily the sole solution to California's water problems. There is simply not enough water to serve all desired uses. The situation surrounding the Delta is a symptom of scarcity. The effective management of scarcity requires not only the best science and technology, but also consideration of public and private values, usually through political processes, to arrive at plans of action which are scientifically sound but also incorporate and reflect the mix of differing societal values.

This review contains a background section describing the geography, hydrology, and history of the Delta and more detailed explications of the points noted above. Then the discussion is organized according to: (1) critical gaps in the scope of the draft BDCP, (2) the use of science in the draft BDCP (3) adaptive management in the BDCP, and, (4) the fragmentation of management that appears to characterize the effort.

2

Background

The BDCP has been developed in an environment characterized by complexity and uncertainty. Furthermore, the BDCP context is dynamic, with underlying conditions themselves in flux. Complexity and uncertainty characterize the biophysical environment, including complexities and changes in the hydrologic system, such as interactions of altered freshwater discharge regimes of tidal influences, changes in the composition and numbers of many species, variability and changes in precipitation, nutrient and sediment input, and changes in the built environment. They also characterize the human environment, particularly with regard to population growth; people's livelihoods and lifestyles; political, financial, and economic conditions; changes in technology; and changes in people's understanding of these systems. Uncertainty is inherent in many of the above factors. The panel did not consider all of the above factors during its review because to do so would be difficult, time-consuming, and beyond the panel's charge. Nevertheless, recognition of the difficult environment in which the BDCP is being developed is helpful in gaining an understanding and appreciation of the difficulties surrounding it and other attempts to improve the reliability of water supplies in California and to restore the Delta ecosystem. The panel thus briefly summarizes the history and the human and biophysical environment of the region.

The San Francisco Bay Delta Estuary (Delta, for short) includes the lower reaches of the two most important rivers in California and the eastern estuary and associated waters of San Francisco Bay. The Sacramento and San Joaquin Rivers and their tributaries include all of the watersheds that drain to and from the great Central Valley of California's interior, as shown in Figure 2. The respective deltas of these rivers merge into a joint delta at the eastern margins of the San Francisco Bay estuary. The Delta proper is a maze of canals and waterways flowing around more than 60 islands that are protected by levees. The islands themselves were historically reclaimed from marshlands as agricultural lands, and most of them are still farmed.

Today, the Delta is among the most modified deltaic systems in the world (Kelley, 1989; Lund et al., 2010). The Sacramento-San Joaquin Delta, as shown in Figure 3, is an integral part of the water supply delivery system of California. Millions of acres of arid and semi-arid farm lands depend upon the Delta for supplies of irrigation water, and approximately 25 million Californians depend upon transport of water through the Delta for their urban water supplies. Population growth anticipated for the first half of the 21st century is likely to create additional water demands despite significant reductions in per capita consump-

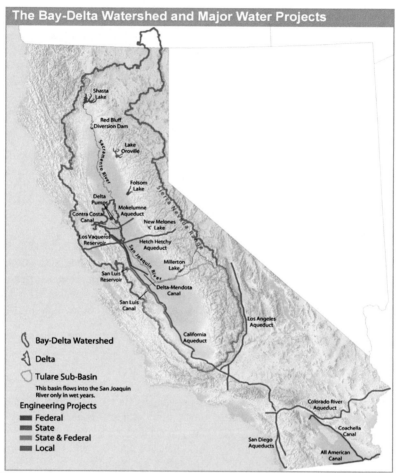

FIGURE 2. The Bay Delta Watershed. SOURCE: Reprinted, with permission, from National Resources Defense Council (*http://www.nrdc.org*).

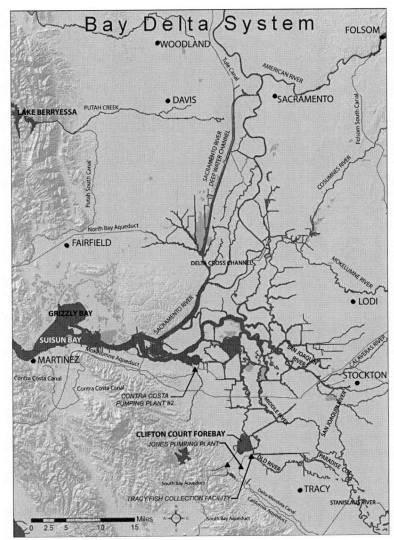

FIGURE 3. The Sacramento-San Joaquin Delta in California. San Francisco Bay, an integral part of the system, is just to the west. SOURCE: Reproduced from NRC (2010b), modified from FWS (2008).

tive uses. In addition, the Delta provides habitat for fish and wildlife, some speciesof which are listed as threatened or endangered under the federal Endangered Species Act and the California Endangered Species Act. The Delta is also an important recreational resource supporting significant boating and fishing activities.

Unimpaired inflows of water to the Delta originate in the watersheds of the Sacramento and San Joaquin Rivers. In an average year those flows are estimated to be 40.3 million acre feet (MAF) or 48.8% of California's average annual total water resource of approximately 82.5 MAF. Of the total unimpaired average inflow, 11.4 MAF are diverted upstream of the Delta for agricultural (83.8%), urban (15.0%), and environmental (1.2%) uses. Diversions from the Delta itself average 6.35 MAF, a little more than a third of all diversions in the Sacramento-San Joaquin system. Diversions from the Delta are dominated by exports to the irrigation service areas of the federal Central Valley Project (CVP) and the State Water Project (SWP) service areas, which include southern portions of the San Francisco Bay Area, the western side of the San Joaquin Valley, and much of southern California. Significant amounts of water are diverted to irrigate Delta lands, and irrigation return flow is discharged into Delta channels. The average yearly outflow from the Delta remaining after the diversions equals 22.55 MAF (Lund et al., 2010).

The quantities of water reported above are for an average year, but hardly any year in California is an "average" water year. Moreover, averages mask the fact that water supplies are highly variable from one year to another. Thus, for example, in the Merced River, which drains the watershed including most of Yosemite National Park and is a tributary of the San Joaquin River, the average annual flow is 1.0 MAF. Yet the low flow of record for the Merced River is 150,000 acre feet, only 15% of the average flow, while the high flow of record is 2.8 MAF, 280% of the average flow. The variability in flows, which is characteristic of all of the state's rivers, is largely a function of the interannual patterns of California's Mediterranean climate, which has a wet and a dry season with precipitation falling mainly in the late fall and winter months. In addition, there is considerable variability in the proportion of the precipitation that falls in the mountains as snow, which adds to the variability of the hydrologic regime.

Until recently, planning for water shortage was based on a five-year dry cycle from the 1930s, or on 1977, the driest year of record. However, recent analyses of potential precipitation resulting from different anticipated climate conditions have changed the criteria employed by the state to project water availability. Despite statewide conservation efforts, which are particularly pronounced in the urban sector, increasing restrictions on diversions have reduced the amount of water available for delivery under the terms of SWP and CVP water supply contracts. These projects, which export water to regions of the state that have experienced persistent water scarcity for many decades, are particularly important features of the California waterscape.

The CVP withdraws water from the Delta and conveys it southward into the San Joaquin Valley through a system of canals built and operated by the federal

Bureau of Reclamation and various water user groups. Most of this water is used for agricultural purposes throughout the San Joaquin Valley and the Tulare sub-basin at the southern end of the Valley. A minor amount is contracted for domestic use. The SWP withdraws water separately from the Delta and conveys it southward to agricultural users on the west side and at the very southern end of the San Joaquin Valley and subsequently over the Tehachapi Mountains into the conurbation of the South Coast Basin. Los Angeles and San Diego are among the water users in the South Coast Basin. The SWP supplies domestic water users in southern California (and a minor amount of domestic use in the southern San Francisco Bay Area) as well as Central Valley agriculture in proportions that are determined in any given year by climatic factors and the availability of alternative sources of supply. Total available supplies have been constrained in recent years by drought and court decisions.

Changes in the hydrologic and physical integrity of the Delta would constrain and threaten the ability of state and federal water managers to continue exporting water in accustomed quantities through the two major projects. This is a concern because the structure of the Delta is changing and will continue to change. Lund et al. (2010) identify several factors that today pose significant threats to the Delta, including: (1) continued subsidence of the agricultural lands on the Delta islands; (2) changing inflows of water to the Delta, which appear to increase flow variability and may skew flows more in the direction of earlier times in the water year in the future; (3) sea level rise that has been occurring over the past 6,000 years and is expected to accelerate in the future; and (4) earthquakes, which threaten the physical integrity of the entire Delta system. There is a long history of efforts to solve these physical problems as well as persistent problems of flood control and water quality (salinity). Salinity intrusion from San Francisco Bay now requires a specific allocation of Delta inflows to repel salinity and to maintain low salinity water at the Delta's western margin. This is done by monitoring and managing the average position of the contour line identifying acceptable levels of salinity, known as "X2". Controlling salinity requires outflow releases from reservoirs that could be used for other demands.

Resolution of these problems is complicated by water scarcity generally and because alternative solutions impose differing degrees of scarcity on different groups of stakeholders. Additional allocation problems arise from a complex system of public and private water rights and contractual obligations to deliver water from the federal CVP and California's SWP. Some of these rights and obligations conflict, and in most years there is insufficient water to support all of them. This underscores the inadequacy of Delta water supplies to meet demands for various consumptive and instream uses as they continue to grow. Surplus water to support any new use or shortfalls in existing uses are unavailable, and any change in the Delta's hydrologic, ecological, or physical elements could reduce supplies further. The risks of change, which could be manifested either by increases in the already substantial intra-seasonal and intra-annual variability or through an absolute reduction in available supplies, underscore the existence of water scarcity and illustrate ways in which such scarcity could be intensified.

In its natural state, the Delta was a highly variable environment. The volume of water inflows changed dramatically from season to season and from year to year. Water quality also varied. In wet periods both salinity and chemical inputs (naturally occurring) were diluted. The species that occupied Delta habitats historically were adapted to accommodate variability in flow, quality, and all of the various factors that they help determine. The history of human development of the Delta, both of land use and water development, is a history of attempts to constrain this environmental variability, to reduce environmental uncertainty and to make the Delta landscape more suitable for farming and as a source of reliable water supplies. A full understanding of the historical pervasiveness and persistence of environmental variability underscores the need to employ adaptive management in devising future management regimes for the Delta (Healey et al., 2008).

The history of water development and conflict in California focuses in part on the Sacramento-San Joaquin Delta. Beginning with the California gold rush in 1848, early settlers sought to hold back the seasonal influx of water and to create agricultural lands. The construction of levees played a central role in this effort, which was threatened in the late 1800s and early 1900s by the movement of hundreds of millions of cubic yards of debris from upstream hydraulic mining that passed through the Delta. There followed throughout the first third of the 1900s further work that helped to stabilize a thriving Delta agriculture (Jackson and Patterson, 1977; Kelley, 1989). The CVP, which began operations in the 1940s (Thompson, 1957), and the SWP of the 1960s required conveyance of water from mainstream river channels through the channels and sloughs of the Delta to the extraction points located in the southern Delta from where water is pumped into the Delta-Mendota Canal (CVP) and the California Aqueduct (SWP) for transport south as illustrated in Figure 4. Once these projects became operational there was a need to control salinity, which became an issue that was decided by the courts (Hundley, 2001; Lund et al., 2010).

Since the beginning of CVP operations, diversions of water to users outside the Delta have been limited to ensure that salinity intrusion does not adversely affect local domestic water diverters in the western margins of the Delta. Additionally, the California's constitution requires that the waters of the state be put to "beneficial use," and this criterion is subject to judicial review and determination. The importance of environmental uses of water has been reflected in many state regulatory decisions and, more recently, in judicial interpretations of the federal Endangered Species Act and the California Endangered Species Act. Several species of Delta fishes and anadromous fishes that migrated through the Delta have been listed as threatened and endangered. The courts became involved, and specific operational restrictions followed from their findings. The maze of federal and state laws as well as the interests of dozens of stakeholder groups have combined to create a gridlock, which at times appeared penetrable

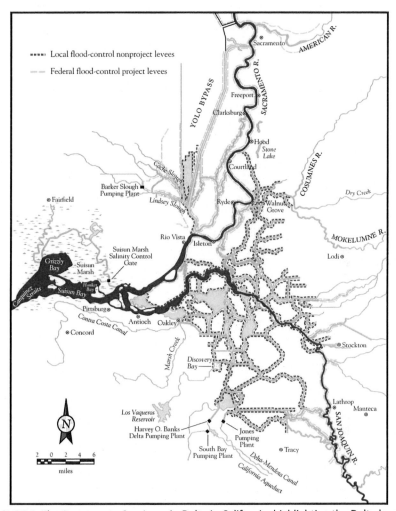

FIGURE 4. The Sacramento-San Joaquin Delta in California, highlighting the Delta levees, 2006. San Francisco Bay, an integral part of the system, is just to the west. SOURCE: Reprinted, with permission, from Lund et al. (2010). Copyright by Public Policy Institute of California.

only by the state and federal courts (Lund et al., 2010). As a result, most recent water operations have tended to be based on legislative requirements and judicial decisions mandating the protection of individual species rather than the optimization of water allocation among all purposes.

There have been several efforts to resolve differences, find areas of agreement, and identify solutions to the problems of the Delta and the operation and allocation of the waters that flow through it. These efforts assumed particular urgency as California was beset by severe droughts in the periods 1987-1992 and again in the first decade of the 2000s. A collaboration of 25 state and federal agencies called the CALFED program was created in 1994 with the mission "to improve California's water supply and the ecological health of the San Francisco Bay/Sacramento-San Joaquin River Delta" (*http://calwater.ca.gov/calfed/about/index.html*). State and federal agencies quickly developed a proposal for water quality standards titled Principles for Agreement on Bay-Delta Standards between the State of California and the Federal Government, otherwise known as the Bay Delta Accord. State and federal agencies with responsibilities in the Delta and stakeholders engaged in a decade-long CALFED process, but they did not alter the strategy of relying on the Delta to convey crucial elements of the water supply to California. The CALFED process also would be used to attain the four main goals of water supply reliability, water quality, ecosystem restoration, and enhancing the reliability of the Delta levees (CALFED, 2000).

The Bay-Delta Accord, which was signed in 2000, began to unravel mid-decade as environmentalists and water users came to believe that their interests were not being well served (Lund et al., 2010) and as federal resources declined. There followed an attempt by the governor to develop a Delta Vision Strategic Plan or "Delta Vision" with the aid of an independent Blue Ribbon Task Force. The Delta Stewardship Plan, which is referred to in this report as the Delta Plan, resulted from this effort. The Delta Plan is a broad umbrella plan mandated by the California Delta Protection Act of 2009 (California Water Code, 85300) to advance the co-equal goals of providing a more reliable water supply for California and protecting, restoring and enhancing the Delta ecosystem. The act requires development and implementation of the plan by January 2012 and specifies that a Delta Stewardship Council, whose membership must reflect broad California water interests, oversee the effort. Also beginning in mid-decade, federal, state, and local water agencies, state and federal fishery management agencies, environmental organizations, and other parties began work on the Bay Delta Conservation Plan (BDCP), a draft of which is the subject of the present report. In addition to the activities already mentioned, many other efforts are ongoing in the Delta such as, for example, a recent report of the State Water Resources Control Board on flows, recent biological opinions concerning listed species, The California Water Plan, The Recovery Plan for Central Valley Salmonids, and the Interim Federal Action Plan.

The BDCP is a habitat conservation plan that can be incorporated into the Delta Plan described above if specific criteria specified in California's water legislation are met (draft BDCP, pp. 1-6). The organizations involved in the

BDCP process have formed a steering committee that includes representatives from the various agencies and interest groups involved in the collaboration (see Appendix B). The BDCP planning effort began in 2006 with a completion goal of 2013. The completed plan also is intended to be implemented over the next 50 years (*http://baydeltaconservationplan.com/Home.aspx*). As of November 22, 2010 close to $150 million has been spent in developing the plan (Sagouspe, 2010).

The BDCP is to be supported by the Environmental Impact Report (EIR)/Environmental Impact Statement (EIS) that will evaluate the range of alternatives for providing ecosystem restoration, water conveyance and other management alternatives identified in the BDCP. The EIR/EIS is currently being prepared by the California Department of Water Resources, the U.S. Bureau of Reclamation, the U.S. Fish and Wildlife Service, and the National Oceanic and Atmospheric Administration's National Marine Fisheries Service in cooperation with California's Department of Fish and Game, the U.S. Environmental Protection Agency, and the U.S. Army Corps of Engineers (*http://baydeltaconservationplan.com/Home.aspx*).

The subsequent sections of this report describe and analyze prominent features of the BDCP while identifying and discussing the critical gaps in the document.

3

Critical Gaps in the Scope of the Draft BDCP

The panel concludes that the draft BDCP is missing critical elements, including an effects analysis, a description of how and where scientific information was used in the draft BDCP, and a description of the BDCP's relationship to other ongoing efforts. In addition, the draft has several structural or systematic problems, including lack of clarity as to the purpose of the BDCP; an unclear linkage of various parts of the BDCP to the effects analysis[4] and among its other components; and lack of detail about analyses of various future scenarios, including a lack of analyses of tradeoffs among the BDCP's goals in various scenarios. The panel offers some guidance on how these systematic problems might be addressed and how the draft BDCP might be completed more usefully.

At the outset of its review the panel identified a problem with the geographical and hydrologic scope of the draft BDCP. The BDCP aims to address management and restoration of the San Francisco Bay Delta Estuary, an estuary that extends from the Central Valley to the mouth of San Francisco Bay. Thus, given that the BDCP purports to describe a *Bay* Delta Conservation Plan, the omission of analyses of the effects of the BDCP efforts on San Francisco Bay (aside from Suisun Bay) is notable. This omission should be of concern to all BDCP parties because the Bay-Delta system is an estuary, and there are significant physical, biogeochemical, and ecological connections between the various sub-embayments as well as between the Bay-Delta and the Pacific Ocean (e.g., Cloern et al., 2010). In particular, changes in outflows and in the tidal prism associated with changing water-project operations and restoration actions would be expected to cause changes in San Francisco Bay, and not only in the Delta. A plan intended to be comprehensive should incorporate these fundamental features of the system. Although the statutory basis of the BDCP may argue against consideration of the effects outside the statutory Delta, the BDCP's failure to address issues related to San Francisco Bay is a significant flaw that should be corrected in subsequent versions of the plan.

---

[4] Even though the effects analysis is not yet complete, the BDCP's authors should at least be able to describe how the completed parts of the BDCP will be linked to the effects analysis.

THE LACK OF AN EFFECTS ANALYSIS

The draft BDCP describes an effects analysis as:

"the principal component of a habitat conservation plan [HCP]. . . . The analysis includes the effects of the proposed project on covered species, including federally and state listed species, and other sensitive species potentially affected by the proposed project. The effects analysis is a systematic, scientific look at the potential impacts of a proposed project on these species and how these species would benefit from conservation actions" (draft BDCP p. 5-2).

Clearly, such an effects analysis, which is in preparation, is intended to be the basis for the choice and details of those conservation actions. Its absence in the BDCP, therefore, is critical gap in the scope of the science and the conservation actions. Nevertheless, the panel presents its vision of the structure and content of a useful effects analysis.

The above description of the effects analysis to be included in the BDCP is rather narrowly cast, because it focuses on the BDCP as a habitat conservation plan (HCP), that is, as an application for an incidental take permit. It thus *presupposes* the choice of the project to be permitted. By contrast, a broadly focused conservation strategy, which the draft BDCP also says it is[5], requires a similarly broadly focused, comprehensive effects analysis. Such an effects analysis would include a systematic analysis of the factors affecting species and ecosystems of concern and the likely contribution of human-caused changes in the system. Such an analysis would then lead to the informed choice of options for reversing the decline of the ecosystem and its components, rather than only analyzing a pre-chosen option. What would such an effects analysis look like?

Effects analyses are used in a range of disciplines to understand complex systems. As noted in the quote above, their main attribute is that they are systematic scientific analysis. Their precise form is not critical. For example, failure mode and effects analysis (FMEA) is commonly applied in the automotive, aerospace, and software industries to understand whether and how the failure of individual components impact the reliability of the overall system (Gilchrest, 1993; McDermott et al., 2009). In the environmental field, effects analyses are used to understand and compare likely responses to alternative management schemes (e.g., Marcot et al., 2001). The National Research Council has reviewed the application of effects analysis within the environmental arena (NRC, 2009). In addition, several NRC reports have discussed or applied the techniques of effects analysis even though they were not necessarily called "effect

---

[5] The following statement appears on p. 1-1 of the draft BDCP: "The [BDCP] sets out a comprehensive conservation strategy for the Delta designed to advance the co-equal planning goals of restoring ecological functions of the Delta and improving water supply reliability to large portions of the state of California."

analysis" (e.g., NRC, 1995, 2002, 2004a, 2004b, 2005; Appendix E of this report provides an example of an effects analysis from NRC 2004b). Effects analyses are commonly used because they integrate empirical data and expert opinion to guide management decisions (e.g., NRC, 2004b). The analytical approaches used in the different types of effects analyses vary from classical risk priority numbers, to simulation modeling (e.g., Legault, 2005), to complex Bayesian network models (Ellison, 1996; Uusitalo, 2007). However, certain important elements are common to all of these analyses, including the need to describe how individual components in the system are connected. It is an effects analysis of this scope that the panel envisions for the BDCP. Here, the panel provides guidance regarding the structure and essential elements that it would expect to see in the completed effects analysis for the draft BDCP. The panel draws on a recent paper by Murphy and Weiland (2011) for a description of a useful effects analysis, itself based to some degree on NRC (2009), because it sets forth specifics for an effects analysis that would be appropriate for the Delta. The panel agrees with Murphy's and Weiland's general approach.

An effects analysis is an essential element of the final BDCP, because it will help meet the legal requirement for a habitat conservation plan to evaluate whether the preferred action aids in the recovery of the species (state requirement) and does not appreciably reduce the likelihood or the survival and recovery of the listed species in the wild (federal requirement). These requirements are initially triggered because as an HCP/NCCP (natural communities conservation plan), the BDCP deals with listed species. However, even if this were not the case, an effects analysis provides the framework within which the impacts of alternative management options can be compared and thus could be justified from a purely logical point of view. An effects analysis is further justified because it also may inform the adaptive management process by identifying which components or processes are the most sensitive indicators of the status and structure of the ecosystem (McCann et al., 2006).

Once the goal of the effects analysis has been defined, the first element of any effects analysis must be an integrated description of the components of the system and how they relate to one another. This description should include a clear statement of the alternative management actions proposed, including that of no action. The activities in this first section naturally lead to a clear definition of the management goal and the temporal and spatial domain of the impacted area. At this introductory level, it is not necessary to quantify the relationships. One needs to mainly indicate the connections. Such a description is essential for several reasons. Most important, it formalizes the understanding of the connections among processes and components in the system. It defines which processes and components are expected to respond to any perturbation and which ones will not. Secondarily, in formulating the problem, a conceptual diagram can serve to identify and rank in importance data on different processes and components within the system. Finally, the system description provides a broader context into which information on the status and trends of species covered by state and federal statues can be placed—such that the dependencies of these listed species on processes and components of the system are identified.

The second stage of the effects analysis should be the collection, review and critical assessment of the best relevant scientific information available. The determination of which data need to be assembled is guided largely by the conceptual framework identified in the first stage. It is neither necessary nor helpful for the assembled data to be encyclopedic in coverage. However, it is essential that data on those processes and components identified in the first stage are compiled, assessed and summarized. This information may be in the form of empirical data or in instances where data are unavailable, in the form of expert opinion. Expert and stakeholder opinion has been successfully used in several management questions involving water use or fish stocks (Borsuk et al., 2001; Miller et al., 2010). The objectives of the data assembly phase are to clearly describe the baseline or reference condition[6] and to quantify the expected relationships among system processes and components. An important feature of this stage is the need to include information on the uncertainties around estimates of processes or component levels. Additionally, the spatial and temporal scale of processes and components under consideration are a vital concern. Different processes and components likely respond at characteristic spatial and temporal scales. For example, the response of many chemical or physical variables might scale with the residence time of water in the system, whereas the response of biological variables might scale with the generation time of the organisms involved. Similarly, salinity gradients affect much of the central and western Delta, while some organisms like salmon, which spend a portion of their life cycles in sea water, occupy much of the North Pacific as well as the Delta and its tributaries. Within the biological realm, rates of primary production, nutrient and oxygen cycling, as well as microbial growth may respond rapidly to ecosystem conditions whereas the abundance of long-lived animals such as sturgeon is expected to integrate ecosystem dynamics over extended periods. The Comprehensive Everglades Restoration Plan (CERP) provides a good example of the use of measurable outcomes for these purposes (NRC, 2008, 2010c).

The next stage of the effects analysis is the most challenging–that of representing the dynamic response of the system. For simple systems, this may be in the form of a simple model. For example, decisions regarding quota levels in fisheries management are often made with guidance from a single assessment model, albeit one with hundreds of parameters (Miller et al., 2010). However, even in simple systems, the level of uncertainty present in individual processes and components of the system may be of such magnitude that state-variable models are unreliable. In these cases probabilistic models have been developed (Legault, 2005). More recently, Bayesian approaches have been used to guide management in the face of uncertainty for complex environmental questions

---

[6] Large restoration programs usually include methods for assessing their effects so that adaptive management can occur. The basic prerequisite for such assessments is the establishment and characterization of a reference condition against which future conditions and proposed alternatives can be compared.

(Borsuk et al., 2004; McCann et al., 2006; Rieman et al., 2001). For an example of incorporating uncertainty into management options, see Box 1.

In the case of the BDCP, it is unlikely that a single analytical framework, even one as flexible as Bayesian network analysis, will be adequate. Thus, it is likely that multiple models will be used to assess the response of different system components to each management alternative. Ultimately a range of integrated scenarios should be developed that link the models' outputs to an integrated response. It is particularly important that each set of the models and analyses be clearly related back to the original conceptual framework generated in the first stage of the effects analysis. Analysts should be explicit about the model inputs and assumptions for each stage of the process. One of the risks of this approach is error propagation, that is, that uncertainty inherent in the forecasts made for one component are not fully carried forward to models of other components.

It would be highly advantageous if outcomes in the effects analysis were quantifiable empirically and could thus become components of the BDCP's Monitoring and Evaluation Program (e.g., NRC, 2000, 2008; Orians and Policansky 2009). As noted above, the CERP has considered and described these issues in considerable detail (NRC, 2008 and references therein). This information, when gathered in the BDCP's Monitoring and Evaluation Program, could then be used to conduct statistical analyses and calibrate models and the modeling framework to inform the adaptive management phase over the decades following implementation of the BDCP actions.

BOX 1
The 2008 Federal Columbia River Power System Biological Opinion

A suitable example of an attempt to incorporate uncertainty is evidenced in the 2008 Federal Columbia River Power System (FCRPS) Biological Opinion (NOAA, 2008) and in the 2010 Supplemental FCRPS Biological Opinion (NOAA, 2010) prepared after the 2008 opinion was voluntarily remanded. The comprehensive analysis in this biological opinion focused on determining the effects of different dam operation alternatives, on key ESA-listed anadromous salmonid populations in the Columbia River Basin. In that analysis, water delivery and dam operation models create conditions that route juvenile salmon through different routes at eight dams in the FCRPS, resulting in net smolt survival downstream of the last dam (Bonneville). Changes in smolt system survival associated with different operation-alternatives are then linked to a broader life-cycle analysis to assess the potential for population level responses to selected management actions.

During the meeting on December 8, 2010, in San Francisco, presenters indicated that the effects analysis that will be included in the BDCP will be only a first step, that is, that it would be iteratively updated as empirical data from the operation of the approved alternatives become available. This approach is certainly compatible with the use of the effects analysis framework as the foundation of the adaptive management framework. If this is indeed how the BDCP developers intend to use the effects analysis, the panel recommends that the final version of the plan articulate a clear vision of how the effects analysis will be updated and how these results will be used to generate the ranges that will be the foundation for subsequent adaptive management.

As an example, much of the recent discussion of changes in the Delta ecosystem has focused on declining planktonic primary production in the Delta and Northern San Francisco Bay (Jassby et al., 2002) as driving food-web changes, notably declines in planktonic grazers (secondary producers), that may underlie to some extent the decline of pelagic fish species like delta and longfin smelt (Baxter et al., 2008). Accordingly, significant elements of the BDCP involve efforts to enhance primary and secondary production through creation of additional tidal wetlands mostly around the edges of the Delta, a plan that strongly echoes CALFED's earlier focus on the creation of shallow water habitat (c.f. Brown, 2003). The bases for this strategy are twofold: (1) in the face of light limitations, shallow water habitats for which the photic zone is a greater fraction of the water column should have higher rates of primary production than deeper waters, e.g., channels (Cloern, 2007); and (2) empirically it is observed that the periodically flooded shallow waters of the Yolo Bypass can support high rates of export of phytoplankton biomass (Schemel et al., 2004).

However, if an effects analysis is indeed "the principal component of a habitat conservation plan" (draft BDCP p. 5-2), then it is difficult to see how these and other conservation strategies described in the BDCP can be scientifically justified before the effects analysis is completed.

THE LACK OF CLARITY AS TO THE BDCP'S PURPOSE

The legal framework underlying the BDCP is extraordinarily complex. In attempting to comply with all relevant laws and regulations, the BDCP's authors have undertaken to develop a habitat conservation plan of great importance, scope, and difficulty. The panel recognizes that the authors face significant challenges and that the BDCP is a work in progress. With these caveats in mind, the panel observes that it would be helpful for the draft BDCP to clarify and place into context a number of legal issues, because their nature and interpretation are closely tied to the BDCP's scientific elements. Any lack of legal clarity makes it difficult for the panel and the public to properly understand, interpret, and review the science of the BDCP.

Ambiguous Role of Co-Equal Goals and Their Relationship to the BDCP

According to the draft BDCP (p. 1-8), it:

"has been prepared as a joint [habitat conservation plan] HCP/ [Natural Communities Conservation Plan] NCCP, which will support the issuance of incidental take authorizations from the US [Fish and Wildlife Service] FWS and [National Marine Fisheries Service] NMFS pursuant to Section 10 of the [federal Endangered Species Act] ESA and take authorizations from the California Department of Fish and Game (DFG) under Section 2835 of the [Natural Communities Conservation Planning Act] NCCPA to the non-federal applicants. The BDCP has also been designed to meet the standards of Section 2081 of the California Endangered Species Act (CESA). The BDCP will further provide the basis for biological assessments (BA) to support the issuance of incidental take authorizations from USFWS and NMFS to [the Bureau of] Reclamation pursuant to Section 7 of the ESA, for its actions in the Delta."

Thus, the BDCP is clearly and specifically an application for the incidental take of listed species as set forth in federal and state statutes.

To apply for an exemption from the § 9 "take"[7] prohibition of the federal Endangered Species Act (ESA), the water users must submit a habitat conservation plan (here, the BDCP) that will minimize and mitigate the harmful impacts of their water usage. HCPs prepared as part of an application for an incidental take permit under federal law are not required to help listed species recover, but they must demonstrate that "the taking will not appreciably reduce the likelihood of the survival and recovery of the species in the wild" (ESA § 10).[8] Under state law, the water users must submit a Natural Community Conservation Plan (NCCP) that, among other things, "aids in the recovery of the species." (Natural Communities Conservation Planning Act [NCCPA], Cal. Fish and Game Code §§ 2800-2835). Neither the ESA nor the NCCPA specifically requires applicants to advance the "co-equal goals."

Despite this, the first paragraph of the draft BDCP (p. 1-1) states that it "sets out a comprehensive conservation strategy for the Delta designed to ad-

---

[7] *Take* means "to harass, harm, pursue, hunt, shoot, wound, kill, trap, capture, or collect, or to attempt to engage in any such conduct." ESA, Section 3, 16 U.S.C. 1532.

Harm, within the statutory definition of "take" has been further defined by regulation: "Harm in the definition of take in the Act means an act which actually kills or injures wildlife. Such act may include significant habitat modification or degradation where it actually kills or injures wildlife by significantly impairing essential behavioral patterns, including breeding, feeding, or sheltering." 50 C.F.R. 17.3.

[8] ESA § 10 also requires successful applicants to demonstrate that (1) " the taking will be incidental [to an otherwise lawful activity]," (2) "the applicant will, to the maximum extent practicable, minimize and mitigate the impacts of such taking," (3) "the applicant will ensure that adequate funding for the plan will be provided," and (4) "[such other measures that the Secretary may require as being necessary or appropriate for purposes of the plan] will be met." 16 USC § 1539(a)(2)(B).

vance the co-equal planning goals of restoring ecological functions of the Delta and improving water supply reliability to large portions of the state of California." This and similar statements throughout the plan make it difficult to understand and evaluate the purposes of HCPs and NCCPs, and the methods of implementing them. Moreover, the methods of implementation are considerably different from the purposes and methods for achieving the two co-equal goals specified in California statutes. Indeed, California has begun to develop a broader "Delta Plan" in accordance with a recent state statute (Cal. Water Code §§ 85300-85309). Thus, the question arises as to the degree of importance to the BDCP of its purpose as an HCP/NCCP and of its purpose as a broader conservation plan designed to achieve California's two co-equal goals. The BDCP and the Delta Plan address the same ecosystem and are somewhat overlapping, but their goals and legal requirements are not identical. Unless the BDCP's relationship to the Delta Plan is clearly described, and its purposes clearly delineated, it will be difficult to assess the BDCP's underlying scientific basis, because the purposes of a broad conservation plan like the Delta Plan are not necessarily the same as those of a habitat conservation plan.

The body of the BDCP contains some elements of both purposes, but not in a coherent and consistent way. For example, despite the statement that achieving the two co-equal goals is one of its purposes, the BDCP focuses on one of the goals at the expense of the other. Additional sources of the confusion are multiple, but two stand out. First, the BDCP document lists some eight planning goals of which providing a "basis for permits necessary to lawfully take covered species" is only one of these eight goals (draft BDCP, p. 1-6). Yet, the remainder of the BDCP appears to focus disproportionately on this goal. As such, much of the BDCP appears to be a post-hoc rationalization of the water supply elements contained in the BDCP.

A consequence of the lack of clarity is related to this post-hoc rationalization. To the extent that the BDCP is simply a request for an incidental take permit then the water users would first identify their desired action (such as construction of a specifically configured "alternative conveyance"), and then analyze its impacts and to develop measures to minimize and mitigate adverse effects. However, to the extent that the BDCP seeks incorporation into the broader Delta Plan, then an effects analysis would precede the choice of all conservation and alternative-operation options, and only then would an effects analysis of those options be performed. That is, if the proposed conveyance system and other measures such as wetlands restoration have been developed as measures to further the restoration of the Delta ecosystem, then one would expect that the effects analysis would be completed before coming to a conclusion as to the preferred type of water delivery system. The absence of an effects analysis and of consideration of water supply alternatives (other than the 45 mile tunnel or possibly an open canal; see section below on alternatives) suggests that the BDCP's major purpose is to provide the basis for an application for an incidental take permit. Yet, this is contrary to what is stated throughout the plan with respect to the attainment of co-equal goals.

Despite these ambiguities, the draft BDCP has concluded that an "isolated conveyance facility" should be constructed consisting of a 45-mile tunnel or pipeline, capable of conveying 15,000 cubic feet per second (cfs) of Sacramento River water around the Delta to the south Delta's existing water export pumping plants, to allow for "dual operation" with the existing south Delta diversion facilities (draft BDCP, Chapter 4.2.2.1.1 and Table 4-1). (Again, the "note to reviewers" on p. 4-14 of the draft BDCP suggests that the conveyance system might be a canal, but there is no analysis of a canal in the draft BDCP or even a statement as to whether the findings from the analysis of a canal would differ from the analysis of a tunnel system.)

Alternative Actions

To support the issuance of an ESA § 10 take permit, the BDCP must specify "what alternative actions to such taking the applicants considered and the reasons why such alternatives are not being utilized" (ESA § 10, 16 U.S.C. § 1539(a)(2)(A)). Even if the proposed action has been decided on, an analysis of alternatives is still required. This analysis does not appear prominently in the draft BDCP. Not only is the analysis a legal requirement, but it also is important scientifically, because to the degree that the reasons for not utilizing the alternatives are *scientific* reasons, the absence of the analysis hinders the ability to evaluate the BDCP's use of science. If the BDCP also seeks incorporation into the Delta Plan (and thereby qualifying for state funding of public benefits), then it should also include an analysis of "conveyance" alternatives. As a prerequisite to incorporation, the BDCP must undertake "a comprehensive review and analysis of . . . [a] reasonable range of Delta conveyance alternatives, including through-Delta, dual conveyance, and isolated conveyance alternatives and including further capacity and design options of a lined canal, an unlined canal, and pipelines" (Cal. Water Code, § 85320). Finally, the federal approval process also will require an environmental impact statement that considers alternatives to the "proposed action," which includes construction of the alternative conveyance (National Environmental Policy Act, 42 U.S.C. § 4332(2)(C)(iii)). Once again, this legally required analysis of alternatives is scientifically important. Therefore, to permit a complete scientific evaluation of the BDCP, it should include an analysis of such alternatives to "take" and to the construction and design of the contemplated isolated conveyance.

4

Use of Science in the BDCP

The panel recognizes the body of scientific information available to support some actions within the BDCP. For example, the compilation of the Delta Regional Ecosystem Restoration Plan (DRERIP, see Appendix F in the draft BDCP) demonstrates that the community has invested considerable effort in establishing a scientific foundation for the numerous actions proposed in the draft BDCP. The participation of 50 analysts and scientists in the construction and scoring of the scientific evaluation worksheets indicates the large effort devoted to identifying ecologically founded actions. The massive DRERIP reflects the collective wisdom and insight of the region's most knowledgeable and respected scientists.

However, it is not clear how the BDCP's authors synthesized the foundation material and systematically incorporated it into the decision-making process that led to the suite of actions selected for implementation. As a unit, the draft BDCP combines a catalog of overwhelming detail with qualitative analyses of many separate actions that often appear disconnected and poorly integrated. Thus, although the biological descriptions and scenarios reflect a strong understanding of the scientific basis for many individual actions by the BDCP authors, there is no obvious distillation, synthesis and integration of the material into a cohesive decision-making process. The BDCP's authors may have performed this critical exercise, but it is not described in the BDCP itself. The panel expects that the pending and critical effects analysis document could provide that convincing clarifying synthesis, relying on the DRERIP to provide the grist. Importantly, the participants who contributed to the DRERIP identified many uncertainties and deficiencies that need to be addressed by the community. Addressing these concerns presumably should happen before the plan is accepted as an ecologically sound path. The following excerpt from the DRERIP emphasizes these points:

> "Collectively, the synthesis team concluded that a number of the conservation measures have the potential for additional synergistic effects that can raise or lower the value of some individual conservation measures when implemented concurrently with other actions. The complexity of various trade-offs between expected positive and negative effects make it difficult to predict the biological responses to concurrent multiple measures. The Synthesis Team recommended that refinements could be made to the proposed modification of the Fremont Weir and Yolo Bypass inundation, North Delta diversions with bypass criteria, and Cache slough restoration to op-

timize ecological benefits and water supply goals. They also identified the
need for better information and modeling of the survival and growth of
covered species and predators to establish baseline conditions against
which benefits can be assessed..." (DRERIP, see Appendix F-1 of the draft
BDCP, p. 17).

This is just one example of the strong body of scientific information that is
available to support specific actions within the plan. Nevertheless, there is a
deficiency in the scientific synthesis that is needed to support the collective actions specified in the BDCP. Some examples of opportunities for demonstrating
that scientific synthesis are described below.

INCORPORATING RISK ANALYSIS

The analyses for the Delta Risk Management Strategy (DRMS, 2009) have
been performed to better understand the various risks to the integrity of levees
and the local and statewide consequences of levee failure. Although there are
limitations to this analysis, the results can offer guidance for prioritizing actions
within the BDCP. For example, the DRMS study indicates that the benefits of
the restorative conservation measures could be lost if levees failed and concludes that current levee management strategies in the Delta are unsustainable
because of seismic risk, high water conditions, sea level rise and land subsidence. In addition to these broad conclusions, the report offers specific estimates
of land impacts (e.g., economic costs of more than $15 billion due to earth-
quake-derived levee failures and associated flooding of 20 islands) (DRMS,
2009).

California continues to invest in levee restoration, and additional restoration
is included in the BDCP. However, levee repairs are not prioritized with regard
to objectives such as habitat restoration, salinity management, drinking water
protection, and preserving agriculture and historic Delta communities. Thus, any
effects analysis should explicitly consider the interactions and tradeoffs between
infrastructure and ecosystem goals. These interactions and tradeoffs may be
considered in a risk-based framework, which could be complemented by analysis of the system reliability (the likelihood that a hydrosystem will fail to
achieve some target), resilience (the ability of a system to accommodate, sur-
vive, and recover from unanticipated perturbation or disturbance), and vulnerability (the severity of the consequences of failure) (Fiering, 1982; Hashimoto,
1982; Moyle et al., 1986).

Furthermore, decision frameworks have recently been demonstrated in the
Delta that highlight the economic tradeoffs of levee repair against the value of
land and assets protected by those levees (Suddeth et al., 2010). The results sug-
gest that, even with doubling of property values, repair of levees is not economi-
cally justifiable for most of the islands within the Delta's Primary
Zone. Although decisions regarding levees, habitat, land use, and water alloca-

tions will certainly be based on more than economic motivations, the use of existing decision analysis tools, and development of new ones to address specific needs, may be invaluable in justifying prioritization of actions and geographical areas of emphasis within and outside of the Delta.

In developing such risk-based approaches, BDCP partners may also identify unacceptable outcomes and evaluate their likelihood, a task that would be valuable in comparing the ability of various management strategies to reduce the likelihood of hydrosystem deterioration, as has been suggested for climate change adaptation (NRC, 2010a). Therefore, the panel recommends that the BDCP partners select and apply a formal analytical framework to investigate the outcomes of proposed activities, including quantitative projections and existing science. Such an analysis—the effects analysis described in some detail above—should occur in advance of selecting the conservation and management actions, and should link specific restoration goals and undesirable system outcomes to the costs, benefits, and reliability of the proposed actions. To do so will require use of the extensive science developed in the basin, recognizing the limitations of its application and the implications of scientific uncertainty in prioritizing actions.

INTEGRATION OF CLIMATE CHANGE ANALYSIS

Climate change has been and will continue to be a major driver of hydrologic and landscape changes in the Delta. Projected changes in the primary drivers of climate change—namely rising temperatures, changing patterns of precipitation, and sea level rise—are expected to result in significant impacts to the ecosystems of both the Delta region and its tributary watersheds and will adversely impact the water supplies that are critical to both urban and agricultural users who depend on the Delta, the major reservoirs and the water conveyance systems (Chung et al., 2009). Therefore, climate change could pose significant threats to the success of the BDCP's ecological goals and could increase the need for additional conservation measures such as construction of additional surface and aquifer storage facilities, demand management such as conservation programs and pricing, and changes in operating strategies (Lund et al., 2010), and it could affect economic factors and water operations (for example, Tanaka et al., 2008).

California's climate change research has generated a wealth of information (Franco et al., 2008), which indicates potential impacts of climate change in the Delta region (e.g., Cayan et al., 2000; Climate Action Team, 2010; DWR, 2010; Field et al., 1999). The work to date has included a systems approach to understanding natural variability including (1) the potential global interconnections to the region's climate (Gershunov et al., 2000; Redmond and Koch, 1991); (2) detection and attribution of historical change in climate (Bonfils et al., 2008); (3) quantification of potential changes in primary stressors of climate through analyses of general circulation model (GCM) predictions (Cayan et al., 2009) and statistical downscaling (Hidalgo et al., 2008; Maurer and Hidalgo, 2008); (4) impacts of projected sea level rise (Knowles, 2008) and effects of rising temperatures on Delta water temperatures (Wagner et al., 2011); and (5) the sensitivity of the water resource system to climate change and sea level rise (USBR, 2008).

Although significant research on climate change vulnerabilities exists in the literature and in various reports produced by numerous agencies and institutions, the panel could not find evidence that such information has been used effectively in the development of the BDCP. Climate change analysis is legally required to obtain an incidental take permit, per NRDC vs. Kempthorne, 506 F.Supp.2d 322 (E.D. Cal. 2007). Yet the draft BDCP's treatment of the topic of climate change, including warming and sea level rise, is fragmented: climate change is addressed only in the descriptions of existing biological conditions (Chapter 2), and sparsely in the Conservation Strategy section (Chapter 3). Furthermore, these discussions are limited to qualitative assessment of potential vulnerabilities and how the conservation strategy might be able to accommodate such impacts. The panel could not find a quantitative analysis of the specific hydrological and biological consequences of potential changes in the primary drivers and consequent changes in the tributary watersheds, aquifers, demands, risks of levee failure, and ecology of the BDCP plan area. Neither could the panel find a statement indicating that such analyses are not available or feasible at this scale. In spite of the brief quantitative summary of potential changes described in Section 2.3.3.2 (pp. 2-36-2-37), there is no evidence that such estimates have been incorporated into the effects analysis and the design of conservation strategy elements. Chapter 5 of the draft BDCP (p. 5-3) says:

> "The effects of climate change (e.g., sea level rise, temperature, and hydrology) were evaluated for early and late points in time of BDCP implementation based on climate change scenarios developed by the consultant team, technical staff from the lead agencies, and outside climate change experts (see Appendix K, *Climate Change Evaluation Methods*, for a discussion of this analysis),"

which appears to address some of the panel's concerns. However, such information was not included in the draft BDCP that was provided.

In the presentation ("Incorporating Climate variability, Change, and Model Uncertainty in Scenarios for California Water Planning") to the panel during its open session on December 8, 2010, Armin Munevar, a consultant from the firm CH2M HILL, did include the aforementioned analysis. A summary of this work appears in a December 2010 report entitled *Climate Change Characterization and Analysis in California Water Resources Planning Studies* (DWR, 2010, pp. 58-67). The climate change study of the BDCP is summarized in the above report and constitutes a reasonable approach for incorporating the current information regarding future climate projections, as predicted by the climate models, and the corresponding hydrologic impacts. Recognizing that precipitation projections are more uncertain (p. 2-36, draft BDCP) than temperature projections, the BDCP's approach includes five scenarios: (1) drier, less warming; (2) drier, more warming; (3) wetter, more warming; (4) wetter, less warming; and (5) a central tendency scenario, which aggregates the majority of model projections (DWR, 2010, p. 62). A further addition to this approach is the concept of the "nearest neighbor" method to select subgroups of models that represent the above five scenarios. Groups of GCM predictions and the corresponding downscaled information demonstrate a significant spread in both precipitation and temperature, and the above approach of using five scenarios to select a set of model runs bracketing the potential changes in precipitation and temperature appears to be adequate until better methods become available.

The above scenarios for climate change and sea level rise have been combined with a variety of hydrologic, operational, and hydrodynamic models to investigate the performance of numerous BDCP scenarios with respect to such metrics as changes in the timing and magnitude of watershed run-off, reservoir storage, flows in the southern part of the Delta, and seasonal variations in the salinity gradient (the position of X2). This analysis appeared to address the hydrologic and hydrodynamic impacts of climate change, incorporating a sequence of linked models to propagate the effects throughout the system.

The panel did not see clear evidence of the use of these hydrologic and hydrodynamic effects to assess the corresponding impacts on ecological processes in the BDCP plan area. According to the DWR 2010 report, the operational simulations of the BDCP using DWR's CALSIM II model have not been completed. Such an analysis is extremely important for investigating the feasibility of meeting future demands associated with the environmental, agricultural, and urban subsystems connected to the greater Bay Delta system. The panel could not find a clear discussion of the extent to which such demands may or may not be met under future climate change scenarios. In addition, there were no quantitative estimates of trade-offs between the co-equal goals of the plan under climate change scenarios, which is discussed below.

Incorporation of the following key elements would strengthen the BDCP's treatment of climate change: (1) Provide a detailed documentation of the approach, analysis, and conclusions, with emphasis on uncertainties and their implications. The lack of discussion in the material provided to the panel of the plan's approach to climate change makes it difficult to more definitively evaluate the scientific basis for climate change projections. (2) Continue efforts to select models with better skills, including models with the ability to reproduce ocean-atmosphere teleconnections, including regime shifts, in the California region (Brekke, 2008, 2009); (3) Quantify the impacts of warming, changes in watershed hydrology and sea level rise on the ecology of the Delta system though the use of ecological models (e.g., CASCaDE, 2010) and quantify the effects on the plan's co-equal goals; and (4) clearly address the role of climate change in the adaptive management strategy. Considering the length of the planning horizon and the importance of climate change to the plan's success, the panel concludes that the BDCP should include a separate chapter on this subject. In view of the importance of the climate change implications in the planning and implementation of the BDCP, the panel recommends that this work be reviewed in detail by an independent expert panel assembled by the Delta Science Program or the Interagency Ecological Program.

A FRAMEWORK FOR LINKING DRIVERS AND EFFECTS

The comprehensive conceptual framework developed by the Interagency Ecological Program related to the drivers of pelagic organism decline in the Delta is an important example of supporting science (Mueller-Solger, 2010). This framework identifies and links, in the context of both ecosystem structure and functioning, the key stressors that help to explain the decline of pelagic organisms. The "drivers of change" (Figure 5) are quantifiable, "suitable for model evaluation" and directly linked to hydrologic, biogeochemical and biotic changes that accompany diversion of freshwater from the Delta and parallel increases in nutrient and other pollutants resulting from upstream anthropogenic activities. This is an example of how the individual components could be functionally and conceptually linked and of how climate-change modeling should be integrated into other aspects of the BDCP, including regime shifts.

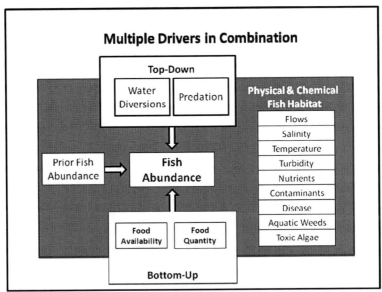

FIGURE 5. Conceptual framework, providing example of supporting science for linking drivers of ecological change to fish community responses. This figure could be a starting point for establishing and rationalizing these linkages. SOURCE: Reprinted, with permission, from Interagency Ecological Program (2010) as modified from Sommer et al. (2007). Available online at: *http://www.water.ca.gov/iep/docs/FinalPOD2010Work plan12610.pdf.*

The types of stressors identified are integrative, reflecting co-occurring physical, chemical, and biotic changes. They also apply to multiple structural (food web structure, biodiversity) and functional (food transfer changes, biogeochemical cycling) changes taking place in the Delta. The framework and associated detail are both comprehensive and useful in terms of linking these drivers to changes taking place at multiple levels of the food web. This type of conceptual approach will also be useful for examining other drivers and impacts of ecological change, including observed changes in fish community structure and production; specifically, how these changes are affected and influenced by changes in physico-chemical factors (e.g., salinity, temperature, turbidity, nutrients/contaminants) and at lower trophic levels (phytoplankton, invertebrate grazers, and prey).

Such a conceptual framework is a necessary precursor to the more holistic integrated analyses for which this panel has identified a need. It may well be impossible to develop a single, integrated model that simultaneously addresses all sources of uncertainty. However, the panel identifies the need for clearer connections among the currently disparate analyses as part of a more synthetic BDCP.

SIGNIFICANT ENVIRONMENTAL FACTORS
AFFECTING LISTED SPECIES

Much of the analysis of factors affecting the decline of smelt and salmonids in the Delta has focused on water operations, in particular, the pumping of water at the south end of the delta for export to other regions. This is in part because the pumping can be shown to kill some fish and in part because proposed changes in water operations were the focus of biological assessments and biological opinions developed by NOAA and USFWS (NRC, 2010b). However, many scientists and others in the region have recognized that other significant environmental factors ("other stressors") have potentially large effects on the listed fishes (e.g., NRC, 2010b). Recent studies have suggested that some of these other factors might be of critical importance to fish (e.g., Baxter et al., 2010; Baxter, 2010; Glibert, 2010). In addition, there remain considerable uncertainties regarding the degree to which different aspects of flow management in the Delta, especially X2 management, affect the survival of the listed fishes (e.g., NRC, 2010b). Indeed, the significance and appropriate criteria for future environmental flow optimization have yet to be established, and are uncertain at best.

The panel supports the concept of a quantitative evaluation of the significance of stressors, ideally using life-cycle models, as part of the BDCP, but such a quantitative evaluation is not part of the draft of the BDCP. The panel concludes that in addition to being incomplete, the absence of a data-based, quantitative assessment and analysis of stressors, ideally using life-cycle models, that supports the effects analysis and adaptive management, is a significant scientific flaw in the current version of the BDCP. A sound, data-supported, quantitative analysis of stressors should be one part of the planning process and should provide the foundation for the effects analysis, adaptive management, and ultimately the choice of conservation measures.

SYNTHESIS

The panel finds the BDCP to be a long list of ecosystem management tactics or incomplete scientific efforts with no clear over-arching strategy to integrate the science, or implement the plan. Furthermore, the BDCP does not tie proposed actions together, in terms of addressing the co-equal goals in a unified way or in terms of ecosystem restoration. On the ecosystem side alone, the plan lists more than 100 restoration actions but provides no guidance on which actions are most important, which actions are more or less feasible, which species are more or less susceptible to extinction, which restoration efforts are most difficult, and which actions might be most easily and immediately addressed. In other words, there is a list but not a synthesized plan for the restoration activities. A systematic and comprehensive plan needs a clearly stated strategic view of what each major component of the plan is trying to accomplish, how it is going to do it, and why it is justified. Also, a systematic and comprehensive plan would show how the co-equal goals are coordinated and integrated into a single resource plan.

THE RELATIONSHIP OF THE BDCP TO
OTHER SCIENTIFIC EFFORTS

A cohesive conservation plan should provide a clear picture of how the different efforts in the Delta fit together. Indeed, such a synthesis could be valuable not only to the BDCP but also to other conservation efforts in the region. As noted above, the BDCP does not provide adequate perspective on how it fits into, for example, the broader Delta Plan, or on how documents such as the Delta Risk Management Strategy fit into the BDCP. Also, aspects of the BDCP fundamental to understanding how and what science was applied are not yet developed. The inadequacies of ingredients such as the effects analysis, or the details of adaptive management or monitoring, lead the panel to ask, how will these tools be employed to assure effective implementation of the BDCP? How specifically will they be tied to the proposals for conservation and infrastructure change? Evidence of a *coordinated* conservation and water management strategy is the first step in establishing public trust that this is a scientifically credible effort.

Clarification of the volume of water to be diverted or mention of how it will be diverted is crucial to a scientific analysis. Moreover, it is unclear how the upper capacity limit of the isolated conveyance structure of 15,000 cfs (draft

BDCP Chapter 4.2.2.1.1 and Table 4-1) was established. The BDCP cannot be properly evaluated if it does not clearly specify the volume of water deliveries whose negative impacts are to be mitigated. The draft BDCP suggests that the water requirements are based on the amount of acreage and crops that contractors have grown, or on the maximum deliveries specified by the SWP contracts—up to 4.173 MAF/year by 2021 (draft BDCP, Chapters 4.3.1 and 5.1). There is no mention that quantities diverted may be constrained by various provisions of California water law, by possible changes in the extent of irrigated agriculture south of the Delta, and by potential changes in cropping patterns fueled by globalizing forces of supply and demand for food. The draft BDCP also fails to identify and integrate demand management actions with other proposed mitigation actions. A conservation plan should address issues of water use efficiency and should account for future trends in other variables that drive the demand for agricultural and urban water supplies. These issues are directly pertinent to the establishment of a water use strategy and they bear importantly on the costs of restoration actions intended to minimize adverse ecological effects. The BDCP's lack of attention to these issues constitutes a significant omission, given the intensifying scarcity of water in California.

In short, synthesis at all levels is a key ingredient in converting a document into a plan. The lack of synthesis constitutes a systemic problem in the draft BDCP. The panel recognizes that the challenge of linking tactics and strategy with a problem this complex is great, but no plan is either complete or likely to point the way toward success without meeting that challenge.

5

Adaptive Management in the BDCP

Adaptive management is a formal, systematic, and rigorous program of learning from the outcomes of management actions, accommodating change, and thereby improving management (Holling, 1978; NRC, 2003). It has been recommended as part of the solution to many environmental problems (e.g., NRC, 2004a), and it is quite appropriately an important part of the draft BDCP. Adaptive management was developed in response to the difficulty of predicting the outcome of management alternatives in natural systems, because of the many uncertainties involved. Current models, typically used for formulating restoration plans, often lack predictive power. Adaptive management, at least in theory, provides resource managers with an iterative strategy to deal with uncertainties and use science, with a heavy emphasis on monitoring, for planning, implementation, and assessment of restoration efforts (Williams et al., 2009). The BDCP has correctly recognized the importance of adaptive management in its various conservation measures and its developers should be commended for emphasizing this aspect of the plan.

Despite numerous attempts to develop and implement adaptive environmental management strategies, many of them have not been successful (Gregory et al., 2006; Walters, 2007). Walters (2007) concluded that most of more than 100 adaptive management efforts worldwide have failed primarily because of institutional problems that include lack of resources necessary for expanded monitoring; unwillingness of decision makers to admit and embrace uncertainties in making policy choices; and lack of leadership in implementation. Thus many issues affecting the successful implementation of adaptive management programs are attributable to the context of how they are applied and not necessarily to the approach itself (Gregory et al., 2006). In addition, the aims of adaptive management often conflict with institutional and political preferences for known and predictable outcomes (e.g., Richardson, 2010) and the uncertain and variable nature of natural systems (e.g. Pine et al., 2009). The high cost of adaptive management, and the large number of factors involved also often hinder its application and success (Lee, 1999; NRC, 2003). Thus, adaptive management, although often recommended, is not a silver bullet and it is not easy, quick, or inexpensive to implement.

In addition to the above difficulties, Doremus (forthcoming) has advocated an analysis of conditions to determine whether adaptive management is an appropriate strategy before it is undertaken. This is good advice, and by implication it could be followed as a method of evaluating existing adaptive management programs. Doremus argues that three conditions favor the use of adaptive management: the existence of information gaps, good prospects for learning at an appropriate time scale compared to management decisions, and opportunities for adjustment. This panel has not performed a formal analysis of the BDCP's situation in regard to these three conditions, and is not aware of any such analysis, but it does draw some preliminary conclusions. Clearly, the first condition (the presence of information gaps) exists, and the second condition (good prospects for learning) seems likely to exist if the program is designed well. The third condition (opportunities for adjustments) is more problematic. There are pressures for management guarantees; for example, the draft BDCP makes clear that one of its aims is a reliable water supply, and Sagouspe (2010) points out that the Planning Agreement that led to the BDCP provides assurances that "no additional restrictions on the use of land, water, or financial resources" beyond the agreed-on amounts will be required without the agreement of the water users (c.f. Richardson, 2010, cited above). Such agreements on their face seem to reduce opportunities for adjustments, although they do not necessarily preclude them altogether.

All of the above considerations lead as well to a reminder of the need for clear goals, cited in many appraisals of adaptive management (e.g., Milon et al., 1998), and this returns the panel to its earlier concern, namely, that the goals of the BDCP are multiple and not clearly integrated with each other. Despite all of the above challenges, there often is no better option for implementing management regimes, and thus the panel concludes that the use of adaptive management is appropriate for the BDCP.

In light of the above, this panel further concludes that the BDCP needs to address these difficult problems and integrate conservation measures into the adaptive management strategy before there can be confidence in the adaptive management program. In addition, an important step in adaptive management that is often given less attention than the others is the need for a mechanism to incorporate the information gained into management decision-making (e.g., NRC, 2003, 2006, 2008). This matter is critical; it also was raised by the Bay Delta Conservation Plan Independent Science Advisors (draft BDCP, Appendix G) and is discussed further below.

In 2009, the BDCP's developers engaged a group of Independent Science Advisors to provide expertise on approaches to adaptive management in the BDCP (draft BDCP, Appendix G-3). Their advice has been incorporated into the adaptive management program presented in Section 3.7 of the draft BDCP. The Independent Science Advisors' report to the BDCP Steering Committee identified key missing elements in the available documentation at the time, including the formal setting of goals based on problems; more effective use of conceptual or simulation models; a properly designed monitoring strategy to evaluate the effectiveness of conservation measures; and more effective assessment, synthesis, and assimilation of information collected during the implementation. Further, their report recommended an adaptive management framework for the BDCP (Bay Delta Conservation Plan Independent Science Advisors' Report on Adaptive Management, 2009, Figure 1, p. 3). The panel concludes that the Independent Science Advisors have provided a logical framework and guidance for the development and implementation of an appropriate adaptive management program for the BDCP.

Much of the information on the adaptive management program is contained in Section 3.7 of the draft BDCP. A brief description of the management of the adaptive management program is presented in Section 7.35. Identification of uncertainties, a critical step in any adaptive management program, is discussed under each of the Conservation Measures (Section 3.4) and adaptive management considerations are shown in Table 3-20, which is part of Section 3.6, Monitoring and Research Program. Because the details of the adaptive management program are fragmented and occur throughout the BDCP without clear linkages of critical components in one section of the document, it is difficult to obtain an overall assessment of the promise of the adaptive management program. The information is not sufficient to demonstrate that the adaptive management plan is properly designed and follows the guidelines provided by the Independent Science Advisors.

Although the adaptive management framework provided by the Independent Science Advisors recommended a logical, stepwise approach for flow of information (Bay Delta Conservation Plan Independent Science Advisors' Report on Adaptive Management, 2009, Figure 1, p. 3), the adaptive management framework shown in Figure 3-63 of the BDCP (also shown in Appendix E of this report) is significantly different and is missing some key elements. It is not clear how the monitoring and "targeted research" programs were designed using goals and objectives, desired outcomes, and performance metrics to select and evaluate steps outlined in the Independent Science Advisors' report. More important, clearly defined uncertainties at various scales starting with the ecosystem level are not presented adequately in the BDCP. In particular, the role of models is not clearly identified in the adaptive management framework, except in Figure 3-63. Box 5b of that figure simply suggests a refinement of models without identifying them. Also, the BDCP does not make clear whether adaptive management applies to broad, ecosystem goals or narrower goals related to specific natural communities or specific conservation measures, or both. Without this distinction and a clear discussion of the role of adaptive management at the ecosystem level, the draft BDCP does not provide assurance that it will successfully use adaptive management to make adjustments during the planning, design, and operational stages of the project.

The Independent Science Advisors correctly pointed out the need for an emphasis on when and where the active versus passive approaches should be used during the design phase. A passive approach is used when the projects are irreversible in nature, as in the case of a dual conveyance facility whereas an active approach involves experiments to test competing hypotheses in cases of significant uncertainties in ecosystem response. The BDCP lacks details of the types of adaptive management approaches and the specifics of the experimental testing that would be conducted to reduce uncertainties. Passive adaptive management is used when there is a high confidence regarding the anticipated ecosystem response, often predicted by reliable models. However, the BDCP does not explicitly rationalize the particular selections in the adaptive management framework, for example, with regard to proposed creation of wetlands, levee restoration, and conveyance options.

The lack of detail about the adaptive management program's details makes evaluating it difficult. Many details of adaptive management are needed to perform a thoughtful review of it, and in some cases, those details emerge only as the plan is implemented. For these reasons, the panel is unable to provide a detailed review of the adaptive management plan at this stage. However, some comments and suggestions are in order.

First, as mentioned above, an adaptive management program requires clear goals. This point often is overlooked. If the project's management goals are not clear, then it will not be evident how to adapt management in the face of new information. The BDCP does not explain how its multiple goals are to be integrated, but the problem goes deeper: some agreed-on goals, such as sustainability of the ecosystem or having a healthy ecosystem, may no longer be acceptable to all parties when they become more specific or when it becomes clear that not all aspects of the ecosystem can be rehabilitated simultaneously. This problem is not unique to the Delta: it affects other large restoration efforts as well, for example, the Everglades (e.g., Milon et al., 1998; NRC, 2010).

Second, adaptive management requires a monitoring program to be in place. The draft BDCP describes its monitoring plan in considerable detail: Table 3-20, which describes the monitoring for effectiveness of conservation actions, runs more than 80 pages, implying a large amount of monitoring activity. However, because there is no effects analysis, it is difficult to evaluate the scientific basis or to justify the appropriateness of individual elements of the monitoring program, elements which clearly should be tied to the results of the effects analysis. In addition, the panel questions the availability of resources necessary to accomplish the all monitoring described in Table 3-20, especially because additional baseline, compliance, and other monitoring also are described in the BDCP as being necessary.

Third, although all of the elements of an adaptive management program are present in the draft BDCP, some of them are not described in detail and some do not appear to be incorporated into the framework in Figure 3-63 (shown in Appendix E of this report). The panel emphasizes again how important it is for a meaningful adaptive management program to be tied to the results of the effects analysis, or at least related to the same issues being addressed by the effects analysis. If it is not, then it is difficult to see how the monitoring and adaptive management program can inform the implementation of the plan and inform decision makers.

The draft BDCP appropriately cites the Independent Science Advisors' Report on Adaptive Management conclusion that:

"the weakest aspect of most adaptive management plans is in the sequence of steps required to link the knowledge gained from implementation monitoring and research and other sources to decisions about whether to continue, modify, or stop actions, refine objectives, or alter monitoring" (draft BDCP, p. 3-577).

This issue has been addressed by NRC reports on the Everglades restoration (e.g., NRC, 2006, 2008), and it is taken seriously by the Comprehensive Everglades Restoration Program. The panel recognizes the difficulty of understanding from the outside how decisions actually are made, and those elements of the BDCP's adaptive management program that require publication of scientific results and provision of the resulting scientific advice to program managers are a good step in that direction. However, a clearer description of the mechanisms that will enable the scientific results to inform management decisions would be helpful.

Details of two other aspects of adaptive management, stakeholder engagement and interagency coordination, are vague. The way that agencies coordinate their activities and that stakeholders participate in the process can have significant consequences. For example, Linkov and his colleagues (Linkov et al., 2006a,b) have described the use of multicriteria decision analysis to enhance adaptive management, and the NRC (2004b) has provided worked examples of such an approach applied to restoring Atlantic salmon in Maine. Those approaches all depend on input from stakeholders. The concepts of a stakeholder committee to receive public input and a "Decision Body" to adjust water operations are too vague and their functions appear to be too limited to provide guidance. The panel recommends that the BDCP take advantage of the literature on this topic—beginning, but not ending, with the material cited above—to inform its processes.

Finally, the importance of action-related triggers related to environmental conditions or the status of covered species is briefly mentioned in the draft BDCP (draft BDCP, Section 3.7.4, pp. 3-586-3-587), but there is no discussion of their importance and role in the adaptive management program and their relation to the effects analysis.

The essence of adaptive management is to identify major uncertainties about the efficacy of policy actions, then to design field tests or management experiments to directly measure efficacy. Such tests can include field evaluation of alternative feedback decision rules that do or do not include thresholds or triggers for action. Initial adaptive management modeling exercises may screen out policies that require triggers by illustrating the challenges associated with uncertainty about the best triggering conditions. In some cases, however, triggers for action can and have been used, often in conjunction with multi-objective structured decision analysis that includes the values and alternatives preferences of the various stakeholders involved (e.g., Karl et al., 2007; Kiker et al., 2008; Miller et al., 2010).

One such example is a recent effort on the Colorado River, where managers are seeking to establish flow releases to control non-native fish below Glen Canyon Dam[9] (Runge et al., 2011). Through the decision-analysis process, objectives were identified (e.g., manage resources to protect tribal sacred sites and spiritual values, maintain and promote local economies and public services, operate within the authority, capabilities, and legal responsibility of the Bureau of Reclamation). In addition, management strategies were evaluated against the objectives, and tradeoffs between strategies were considered. The process identifies specific triggers (e.g. following High-Flow experimental floods, abundance of native or introduced fish species, flow and sediment load) for management actions (e.g., removal of non-native species, fine sediment slurry, release of stranding flows), while other actions (e.g., mechanical or chemical disruption of fish spawning areas, augmentation of fine sediment) are recommended without triggers. The value of triggers is in the efficiency of managing the system, minimizing expensive actions to when and where they are thought to be necessary for and beneficial to species recovery. Such triggers also would help to design a more-focused monitoring program. However, the challenge of using triggers is in the uncertainty in establishing thresholds for triggering actions. Thus, (Runge et al., 2011) caution that their results do not provide the final decision but instead provide guidance for further consultation by the decision makers. That consultation is likely to require experimentation, modeling, and continued adaptive management.

In summary, the BDCP's adaptive management program is not fully developed. In addition, there remain significant scientific, policy, and management uncertainties about the BDCP's purpose and organization. The panel concludes that the BDCP's developers can benefit significantly from experiences in adaptive management attempted in other large-scale ecosystem restoration efforts. One such example is the CERP, where adaptive management has been a key component since its inception in 1999 (USACE & SFWMD, 1999). As recognized by the NRC (2006), the CERP adaptive management strategy provides a sound organizational model for the execution of a passive approach. More recent activities also include examples of active approaches where field tests have played a major role in the early phases of selected projects (RECOVER, 2010). Key components of the CERP adaptive management program are:

- CERP Adaptive Management Strategy (RECOVER, 2006a);
- Monitoring and Assessment Plan and an Assessment Strategy designed to monitor system-wide responses to determine how well CERP is achieving its goals (RECOVER, 2004; 2006a,b; 2009); and

---

[9] The panel provides this example as a good use of action-related triggers. The success of adaptive management in Glen Canyon in general has been questioned (Susskind et al., 2010).

• CERP Adaptive Management Integration Guide (available in draft form) (RECOVER, 2010).

The above documents detail more than five years of progress in implementing adaptive management in the CERP. The CERP's program includes nine activities, which have been effectively integrated into the standard practice of project planning and life-cycle analysis (NRC, 2006). The integration guide describes how to apply adaptive management concepts to the CERP program and related projects through the identification of key uncertainties and the incorporation of activities into the existing CERP planning and implementation process. Even a soundly implemented adaptive management program is not a guarantee of a successful restoration effort, however. As described in several NRC reports and other documents, several factors outside the purview of the adaptive-management teams and even the program managers have hindered restoration progress in the Everglades. They include financial, political, bureaucratic, legal and other obstacles (e.g., NRC 2006, 2008, 2010), factors certain to influence the implementation of the BDCP as well. But a well-designed and implemented program should improve the likelihood of success in implementing the BDCP.

6

Management Fragmentation and a Lack of Coherence

The management of any science-based process has profound impact on the use of science and adaptive management within that process. The panel was charged with evaluating the use of science and adaptive management, and therefore management of the enterprise falls appropriately within this charge. The absence of any synthesis in the draft BDCP draws attention to the fragmented system of management under which it was prepared—a management system that lacks coordination among entities and clear accountability. No one public agency, stakeholder group, or individual has been accountable for the coherence, thoroughness, and scientific integrity of the final product. Rather, the plan appears to reflect the differing perspectives of federal, state and local agencies, and the many stakeholder groups involved, as noted in the introduction to this report. This is not strictly a scientific issue, but fragmented management is a significant impediment to the use and inclusion of coherent science in future iterations of the BDCP. Different science bears on the missions of the various public agencies; different stakeholders put differing degrees of emphasis on specific pieces of science; and different geographical entities require different kinds of science. The panel concludes that without more coherent and unified, the BDCP's final product, like the current draft, will rely on bits and pieces of science that are not well integrated. Moreover, the lack of coherence in the management of the preparation of the BDCP helps to explain the fragmentation of science and the lack of synthesis.

The discussion of the implementation structure in Chapter 7 of the draft BDCP suggests that the fragmented management that characterizes the preparation of the draft plan is also likely to be a feature of the implementation of the plan that finally emerges. The appointment of a single program manager and creation of an Implementation Office, as envisioned in the draft BDCP, are unlikely—even taken together—to result in a well-integrated, coherent implementation program. The public agencies that are involved in the planning and implementation of the BDCP are a mix of operating and regulatory state and federal agencies. Moreover, their interests are intertwined with those of the stakeholder groups, most obviously water-using and environmental groups. These agencies and stakeholders have differing missions and agendas that are almost certain to conflict from time to time and yet the BDCP has no formal mechanism to deal with such conflicts.

Indeed, the BDCP appears to carve out territorial boundaries that make fragmented, and even perhaps antagonistic, management of the plan's implementation more likely. Thus, for example, the BDCP states, "The [Implementation Office] will not be involved in the development or operation of the [State Water Project] and/or [Central Valley Project] facilities" (draft BDCP, p. 7-5). Further, the plan states, "No general delegation of authority by [the California Department of Water Resources] or the [Bureau of] Reclamation to the Program Manager or one of their employees assigned to the [Implementation Office] will occur" (draft BDCP p. 7-7). The plan also proposes that agency personnel be assigned to populate various BDCP implementation committees. This seems to further ensure that inter-agency conflicts and traditional turf battles will be strongly internalized in the management arrangements. The plan, then, envisions that traditional agency missions and turf will be protected, leaving the program manager to navigate through a maze of conflicting interests without any real authority or capacity to resolve conflicts and otherwise ensure that the management approach is integrated.

There is an important literature on the problem of management fragmentation in the planning and operations management of large water schemes (Conca, 2005; Feldman, 2011; Scholz and Siftel, 2005). There is additional helpful literature on network governance (Kettl and Goldsmith, 2004) and collaborative federalism (Emerson and Murchie, 2010). This work underscores the importance of collaboration, the sharing of authority and power, and acknowledgment of the interests of all stakeholders if the large-scale management of water is to be integrated and successful. The panel recommends that the BDCP's authors give this matter careful attention.

Development and implementation of large restoration and conservation programs such as the BDCP often require a complex structure to incorporate technical, political, and legal realities and the evolving dynamics of both the physical and organizational environments. The panel recommends that the agencies responsible for implementing the BDCP review other examples of large scale restoration programs that have been developed and implemented. One such example is the Boston Harbor Islands National Recreation Area where management coordinates through a General Management Plan executed with several cooperative agreements. Although CalFed dissolved, the former CalFed institutional structure dealt with some of the same management issues. The CalFed experience and associated body of literature could be a useful source of positive and negative lessons.

Another example is the Everglades restoration program (CERP; www.evergladesplan.org), with which several committees of the National Research Council have been involved for many years (NRC, 2006, 2008, 2010c). Since its authorization in the Water Resources Development Act of 2000, the CERP has necessitated the development of a number of coordination processes, agreements, and carefully designed planning and implementation efforts (Figure 6 in Box 2 of this report) to incorporate the unprecedented scope and complexity of the final plan, regulations of the federal and state governments, and stakeholder interests. However, unlike the BDCP, the CERP's focus was more on ecosystem restoration than on concerns about endangered and threatened species.

Unlike the seemingly fragmented structure for the BDCP implementation, the authority for implementing the Everglades program lies with both federal and state agencies with a carefully designed planning process and inter-agency agreements in each step. The Everglades management system has accountability in that the federal and state agencies have a formal agreement on cost-sharing of the entire restoration program and the authority to execute the restoration plan. Furthermore, they have coordination mechanisms, such as the South Florida Ecosystem Restoration Task Force which is a coordination mechanism for many entities involved in the restoration. Specifically, the U.S. Army Corps of Engineers (USACE) and the U.S. Department of the Interior (DOI), in partnership with the lead state agency, the South Florida Water Management District (SFWMD), are responsible for undertaking the CERP's implementation. A continuously evolving Integrated Delivery Plan sets the priority projects that must be implemented. Central to the planning and implementation of a particular project is the Project Implementation Report (PIR) developed by a Project Delivery Team, which constitutes a multi-agency team with strong stakeholder participation (Box 2). Active participation by all agencies with authority and pre-approved CERP Guidance Memoranda (CGMs) ensure agreement on the plan, scientific basis, and the expected benefits in the PIR before it is submitted for approval and authorization for funding (see Figure 3-3 of NRC, 2006). The PIR includes an evaluation of alternative designs and operations for environmental benefits, the costs, and the engineering feasibility (NRC, 2006). Once a project

---

BOX 2
Implementation of Everglades Restoration:
Structure for Inter-agency Collaboration and Stakeholder Involvement

The U.S. Army Corps of Engineers (USACE), Department of the Interior (DOI), and the South Florida Water Management District (SFWMD) are currently implementing a planning process that provides significant opportunity for local, state, federal and tribal governments, as well as public and non-governmental stakeholders to participate in the projects that are being designed and implemented. For each project, an interagency, interdisciplinary Project Delivery Team (PDT) is established. The PDT is led by the USACE and SFWMD Project Managers and includes members from various local, state, federal and tribal governments. Figure 6 illustrates the typical composition and entities that provide input and feedback to the PDTs. Although much work is accomplished in a PDT, additional agency stakeholder and public in-

box continues

BOX 2 Continued

put are received at scheduled points in the planning process. Specifically, such advice is sought as development of project objectives, identification of performance measures, selection of evaluation models, and development and evaluation of alternative plans. Additional opportunities for governmental agencies, stakeholders, and the public to provide input and feedback during the planning process are provided at publicly noticed meetings of the following established groups (a) Governing Board of the SFWMD; (b) South Florida Ecosystem Restoration Task Force (SFERTF); (c) South Florida Ecosystem Restoration Working Group; and (d) The Water Resources Advisory Commission (WRAC).

To ensure that the development and implementation of CERP is based on the best and most recent science available, and to ensure that the restoration program is implemented with an adaptive management approach, a multiagency, multidisciplinary science team called RECOVER has been formed. In addition, the USACE and SFWMD have established an Interagency Modeling Center (IMC) to function as a single point of service for the modeling needs of CERP. As the primary organization responsible for regional and sub-regional modeling for CERP modeling, the IMC conducts system-wide evaluations of CERP implementation plans and updates, and provides modeling support for PDTs.

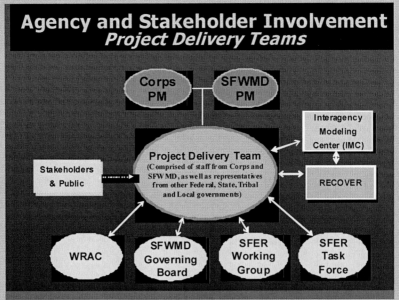

FIGURE 6. Agency and stakeholder involvement in the project delivery teams (PDT). Figure courtesy of the South Florida Water Management District

is authorized, depending on the funding, a series of technical refinements beginning with detailed designs and ending with construction occurs prior to its operation. Project Cooperation Agreements between the federal and the state partner are obtained prior to the initiation of construction. The current progress of CERP has demonstrated the need for formal agreement among partners. One example of such as agreement is the Design Agreement between the USACE and SFWMD (*http://www.evergladesplan.org*). Implementation of the agreement is ensured by an interagency unit known as the Design Coordination Team (DCT), which oversees the schedules and budgets, plans and specifications, and contractual work.

However, no matter how good the management structure may be, it is no guarantee of progress; it is a necessary but not a sufficient condition. Experience with large restoration projects elsewhere, and especially in the Delta, reveals that progress will be affected by lawsuits, economic crises, unexpected (and expected) environmental events, cost overruns, political changes, and so on. Yet the literature and examples mentioned here show that management of complicated systems, where more than one agency has management responsibilities, can be successful as long as there is adequate coordination and clear accountability. Apparently, the new deputy secretary of the California Natural Resources Agency has the BDCP as his major responsibility, which is an encouraging development. The panel recommends that the BDCP's authors give this matter careful attention, because an appropriate system of management is necessary but not sufficient for the use of coherent, synthesized science in future iterations of the BDCP and a successful adaptive management program.

7

In Conclusion

The panel finds the draft BDCP to be incomplete or unclear in a variety of ways and places. The plan is missing the type of structure usually associated with current planning methods in which the goals and objectives are specified, alternative measures for achieving the objectives are introduced and analyzed, and a course of action is identified based on analytical optimization of economic, social, and environmental factors. The lack of an appropriate structure creates the impression that the entire effort is little more than a post-hoc rationalization of a previously selected group of facilities, including an isolated conveyance facility, and other measures for achieving goals and objectives that are not clearly specified. Furthermore, unless goals are not only stated but also prioritized, it is impossible to forecast the effects of projects that would achieve the goals because it is impossible to identify the projects or the consequences that would be deemed acceptable. One symptom of the absence of appropriate structure is the systemic lack of synthesis in the BDCP. Frequently, the plan appears to be little more than a list of tactics or management options that are not strategically integrated. It is unclear how these tactics would be knitted together to achieve the objectives of the plan which are themselves not always clear; and there is no indication of how the various tactics and elements in the plan could be implemented in a logical and strategic fashion.

Several errors of omission also complicate this review. First, there is no effects analysis that describes the impacts of the proposed project or alternatives on target species, even though the BDCP notes that the effects analysis would be "...the principal component of a habitat conservation plan..." Without an effects analysis it is exceedingly difficult to evaluate alternative mitigation and conservation actions. In addition, the plan remains silent on the probable effects of proposed actions on target species. Second, the descriptions of the BDCP's purpose lack clarity. The confusion arises because it is unclear to what extent or whether the BDCP is exclusively a habitat conservation plan, which is to be used as an application for a permit to "take" listed species incidentally, or to what extent or whether it is also intended to be a plan to achieve the co-equal goals of providing reliable water supply and protecting and enhancing the Delta ecosystem. Third, the proposed adaptive management plan is incomplete. Any adaptive management plan requires a monitoring program and, although one is described, it is unclear what purposes it is intended to achieve. The proposed monitoring program has not been linked to the adaptive management plans in a way that would allow managers to account for lessons learned from previous experience, and more important, it is not linked to the effects analysis. In short, there is no compelling information that would allow the panel to conclude that the adaptive management program has been properly designed.

The lack of integrated management and coherence in developing the BDCP is also a shortcoming. The plan reflects the perspectives of various public agencies at the federal, state, and local levels and the many stakeholder groups involved. Although this is not strictly a scientific issue, the panel concluded that fragmented management is a significant impediment to the use and inclusion of coherent science in future iterations of the BDCP. Moreover, the proposed BDCP implementation arrangements appear unlikely to result in a well-integrated, coherent implementation program because of the conflicting agency and stakeholder interests and objectives that are built into the structure of the proposed Implementation Office.

The panel underscores the importance of a credible and a robust BDCP in addressing the various water management problems that beset the Delta. A stronger, more complete, and more scientifically credible BDCP that effectively integrates and utilizes science could indeed pave the way toward the next generation of solutions to California's chronic water problems.

References

Baxter, R. 2010. Are Juvenile Longfin Smelt Abandoning the Suisun Bay Neighborhood? ...The Rest of the Story. Abstract. 6th Biennial Bay-Delta Science Conference. Ecosystem Sustainability: Focusing on Science and Managing California's Water Future. September 8, 2010.

Baxter, R. R. Breuer, L. Brown, M. Chotkowski, F. Feyrer, M. Gingras, B. Herbold, A. Mueller-Solger, M. Nobriga, T. Sommer, and K. Souza. 2008. Pelagic Organism Decline Project Report: 2007 Synthesis of Results. Sacramento, CA: Interagency Ecological Program for the San Francisco Estuary.

Baxter, R., R. Breuer, L. Brown, L. Conrad, F. Feyrer, S. Fong, K. Gehrts, L. Grimaldo, B. Herbold, P. Hrodey, A. Mueller-Solger, T. Sommer, and K. Souza. 2010. Interagency Ecological Program, 2010 Pelagic Organism Decline Work Plan and Synthesis of Results. Sacramento, CA: Interagency Ecological Program for the San Francisco Estuary. Available online at: *http://www.water.ca.gov/iep/docs/FinalPOD2010Workplan12610.pdf.* Last accessed March 10, 2011.

BDCP (Bay Delta Conservation Plan Steering Committee). 2010. Bay Delta Conservation Plan Working Draft. November 18. Available online at: *http://www.resources.ca.gov/bdcp/.* Last accessed April 26, 2011.

Bonfils, C., B.D. Santer, D.W. Pierce, H.G. Hidalgo, G. Bala, T. Das, T.P. Barnett, D.R. Cayan, C. Doutriaux, A.W. Wood, A. Mirin, and T. Nozaw. 2008. Detection and attribution of temperature changes in the mountainous western United States. Journal of Climate 21: 6404–6424.

Borsuk, M.E., R. Clemen, L. Maquire, and K.H. Reckhow. 2001. Stakeholder values and scientific modeling in the Neuse River watershed. Group Decision and Negotiation 10: 355–373.

Borsuk, M.E., C.A. Stow, and K.H. Reckhow. 2004. A Bayesian network of eutrophication models for synthesis, prediction, and uncertainty analysis. Ecological Modelling 173: 219–239.

Brekke, L.D., E.P. Maurer, J.D. Anderson, M.D. Dettinger, E.S. Townsley, A. Harrison, and T. Pruitt. 2009. Assessing reservoir operations risk under climate change. Water Resources Research 45:W04411.

Brekke, L.D., M.D Dettinger., E.P. Maurer, and M. Anderson. 2008. Significance of model credibility in estimating climate projection distributions for regional hydroclimatological risk assessments. Climate Change 89:371–394.

Brown, L.R. 2003. Will tidal wetland restoration enhance populations of native fishes? San Francisco and Watershed Science 1. Available online at: *http://repositories.cdlib.org/jmie/sfews/vol1/iss1/art2.* Last accessed April

13, 2011.

CALFED. 2000. California's Water Future: Framework for Action. Sacramento, CA: CALFED Bay Delta Program.

CASCaDE. 2010. Computational Assessments of Scenarios of Change for the Delta Ecosystem. Available online at: *http://cascade.wr.usgs.gov/index.shtm.* Last accessed March 10, 2011.

Cayan, D., M. Tyree, M. Dettinger, H. Hidalgo, T. Das, E. Maurer, P. Bromirski, N. Graham, and R. Flick. 2009. Climate change scenarios and sea level rise estimates for the California 2008 climate change scenarios assessment. CEC-500-2009-014-D. Sacramento, CA: California Climate Change Center.

Chung, F.I., J. Anderson, S. Aurora, M. Ejeta, J. Galef, T. Kadir, K. Kao, A. Olson, C. Quan, E. Reyes, M. Roos, S. Seneviratne, J. Wang and H. Yin. 2009. Using Future Climate Projections to Support water Resources Decision Making in California. California Energy Commission, PIER Program Report, CEC-500-2009-052-D. Sacramento, CA: California Energy Commission.

Climate Action Team. 2010. Climate Action Team Report to Governor Schwarzenegger and the California Legislature. California Environmental Protection Agency. April 2010. Available online at: *http://www.energy.ca.gov/2010publications/CAT-1000-2010-005/CAT-1000-2010-005.PDF.* Last accessed March 10, 2011.

Cloern, J.E. 2007. Habitat Connectivity and ecosystem productivity: implications from a simple model. American Naturalist 169: E21–E33.

Cloern, J.E., K.A. Hieb, T. Jacobson, B. Sansó, E. Di Lorenzo, M.T. Stacey, J.L. Largier, W. Meiring, W.T. Peterson, T.M. Powell, M. Winder, and A.D. Jassby. 2010. Biological communities in San Francisco Bay track large scale climate forcing over the north Pacific. Geophysical Letters 37, L21602, doi.1029/2010GLO447744.

Conca, K. 2005. Growth and fragmentation in expert networks: The elusive quest for integrated water resources management. Pp. 432–470 in Handbook of Global Environmental Politics, P. Dauvergne, ed. Cheltenham, UK: Edward Elgar Publishers.

Delta Risk Management Strategy (DRMS). 2009. Executive Summary, Phase 1 Report. Available online at: *http://www.water.ca.gov/floodmgmt/dsmo/sab/drmsp/docs/drms_execsum_ph1_final_low.pdf.* Last accessed March 10, 2011.

Doremus, H. Forthcoming. Adaptive management as an information problem. North Carolina Law Review. UC Berkeley Public Law Research Paper No. 1744426. Available online at: *http://ssrn.com/abstract=1744426.* Last accessed on April 11, 2011.

DWR (California Department of Water Resources). 2010. Climate Change Characterization and Analysis in California Water Resources Planning Studies. Sacramento, CA: California Department of Water Resources.

Ellison, A.M. 1996. An introduction to Bayesian inference for ecological research and environmental decision-making. Ecological Applications 6:

1036–1046.

Emerson, K., and P. Murchie. 2010. Collaborative governance and climate change. Pp. 141–161 in The Future of Public Administration, Public Management and Public Service Around the World: The Minnowbrook Perspective, R. O'Leary, S. Kim, D. Van Slyke, eds. Washington, D.C.: Georgetown University Press.

Feldman, D.L. 2011. Water: A Volume in the Geopolitics of Resources Series. Boston, MA: Polity Books.

Field, C. B., G. C. Daily, F. W. Davis, S. Gaines, P. A. Matson, J. Melack, and N. L. Miller. 1999. Confronting Climate Change in California: Ecological Impacts on the Golden State. Cambridge, MA: Union of Concerned Scientists and Washington, DC: Ecological Society of America.

Fiering, M. 1982. A screening model to quantify resilience. Water Resources Research 18(1): 27–32.

Franco, G., D.R. Cayan, A. Luers, M. Hanemann, and B. Croes. 2008. Linking climate change science with policy in California. Climatic Change 87(Suppl 1):S7–S20.

FWS (Fish and Wildlife Service). 2008. Biological Opinion on Coordinated Operations of the Central Valley Project and State Water Project. Available online at: *http:www.fws.gov/sacramento/ed/documents/SWP-CVP_OPs_ BO_12-15_final_OCR.pdf.* Last accessed March 15, 2010.

Gershunov, A., T.P. Barnett, D. R. Cayan, T. Tubbs, and L. Goddard. 2000. Predicting and downscaling ENSO impacts on intraseasonal precipitation statistics in California: The 1997/98 Event. Journal of Hydrometeorology 1:201–210.

Glibert, P.M. 2010. Long-term changes in nutrient loading and stoichiometry and their relationships with changes in the food web and dominant pelagic fish species in the San Francisco Estuary, California. Reviews in Fisheries Science 18 (2):211–232.

Gilchrest W. 1993. Modeling failure modes and effects analysis. International Journal of Quality and Reliability Management 10(5):16–23.

Gregory, R., D. Ohlson, and J. Arvai. 2006. Deconstructing adaptive management: criteria for applications to environmental management. Ecological Applications 16(6):2411–2425.

Hashimoto, T. 1982. Reliability, resiliency, and vulnerability criteria for water resource system performance evaluation. Water Resources Research 18(1):14-20. Available online at: *http://www.climatechange.ca.gov/climate_action_team/reports/index.html.* Last Accessed on April 11, 2011.

Healey, M., M. Dettinger, and R. Norgaard. 2008. The State of Bay-Delta Science, 2008: Summary. Available online at: *http://www.science.calwater.ca.gov/publications/sbds.html.* Last accessed on April 11, 2011.

Holling, C.S. 1978. Adaptive Environmental Assessment and Management. Chichester, UK: John Wiley and Sons, Ltd.

Hundley, N., Jr. 2001. The Great Thirst. Californians and Water: A History. Revised Edition. Berkeley, CA: University of California Press.

Independent Science Advisors. 2009. Bay Delta Conservation Plan Independent Science Advisors' Report on Management, prepared for the BDCP Steering Committee, February 2009. Available at: *http://baydeltaconservation-plan.com/Libraries/Background_Documents/BDCP_Adaptive_Management _ISA_report_Final.sflb.ash.* Last accessed on April 11, 2011.

Interagency Ecological Program (IEP). 2010. Interagency Ecological Program 2010 Pelagic Organism Decline Work Plan 2. Available online at: *http://www.water.ca.gov/iep/docs/FinalPOD2010Workplan12610.pdf.* Last accessed April 25, 2011.

Jackson, W. T., and A. M. Patterson. 1977. The Sacramento-San Joaquin Delta: The Evolution and Implementation of Water Policy, An Historical Perspective. California Water Resources Center Contribution no. 163. Davis, CA: California Water Resources Center. Available online at: *http://escholar-ship.org/uc/item/36q1p0vj#page-20.* Last accessed April 25, 2011.

Jassby, A.D., J.E. Cloern, and B.E. Cole. 2002. Annual primary production: Patterns and mechanisms of change in a nutrient-rich tidal ecosystem. Limnology and Oceanography 2002. 47:712.

Karl, H.A., L.E. Susskind, and K.H. Wallace. 2007. A dialogue not a diatribe—effective integration of science and policy through joint fact finding. Environment 49(1):20–34.

Kelley, R. 1989. Battling the Inland Sea. Berkeley, CA: University of California Press.

Kettl, D. F., and S. Goldsmith, eds. 2009. Unlocking the Power of Networks: Keys to High Performance Governance. Washington, D.C.: Brookings Institution Press.

Kiker, G. A., T. S. Bridges, and J. Kim. 2008. Integrating comparative risk assessment with multi-criteria decision analysis to manage contaminated sediments: an example for the New York/New Jersey harbor. Human and Ecological Risk Assessment 14(3):495–511.

Knowles, N. 2008. Potential Inundation Due to Rising Sea Levels in the San Francisco Bay Region. California Climate Change Center. CEC-500-2009-023-F, California Energy Commission, PIER Energy-Related Environmental Research. Available online at: *http://www.energy.ca.gov/2009publi-cations/CEC-500-2009-023/CEC-500-2009-023-D.PDF.* Last accessed March 10, 2011.

Lee, K.N. 1999. Appraising adaptive management. Conservation Ecology 3(2):3. Available online at: *http://www.consecol.org/vol3/iss2/art3.* Last accessed on April 11, 2011.

Legault, C.M. 2005. Population viability analysis of Atlantic salmon in Maine, USA. Transactions of the American Fisheries Society 134:549–562.

Linkov, I., F.K. Satterstrom, G. Kiker, C. Batchelor, T. Bridges, and E. Ferguson. 2006a. From comparative risk assessment to multi-criteria decision analysis and adaptive management: Recent developments and applications. Environment International 32:1072–1093.

Linkov, I., F. K. Satterstrom, G. Kiker, T.S. Bridges, S.L Benjamin, and D.A.

Belluck. 2006b. From optimization to adaptation: Shifting paradigms in environmental management and their application to remedial decisions. Integrated Environmental Assessment and Management 2(1):92–98.

Lund, J., E. Hanak, W. Fleenor, W. Bennett, R. Howitt, J. Mount, and P. Moyle. 2010. Comparing Futures for the Sacramento–San Joaquin Delta. Berkeley, CA: University of California Press.

Marcot, B.G., R.S. Holthausen, M.G. Raphael, M.M. Rowland, and M.J. Wisdom. 2001. Using Bayesian belief networks to evaluate fish and wildlife population viability under land management alternatives from an environmental impact statement. Forest Ecology and Management 153: 29–42.

Maurer, E. P., and H. G. Hidalgo. 2008. Utility of daily vs. monthly large-scale climate data: An intercomparison of two statistical downscaling methods. Hydrology and Earth System Sciences 12:551–563.

McCann, R.K., B.G. Marcot, and R. Ellis. 2006. Bayesian belief networks: applications in ecology and natural resource management. Canadian Journal of Forest Research-Revue Canadienne De Recherche Forestiere 36: 3053–3062

McDermott, R.E., R.J. Mikulak, and M.R. Beauregard. 2009. The Basics of FMEA. New York, NY: CRC Press

Miller, T. J., J. A. Blair, T. F. Ihde. R. M. Jones, D. H. Secor and M. J. Wilberg. 2010. FishSmart: an innovative role for science in stakeholder-centered approaches to fisheries management. Fisheries 35(9):424–433

Milon, W,J, C.F. Kiker, and D.L. Lee. 1998. Adaptive ecosystem management and the Florida Everglades: more than trial-and-error? Water Resources Update 113:37–46.

Moyle, P.B., H.W Li, and B.A. Barton. 1986. The Frankenstein effect: impact of introduced fishes on native fishes in North America. Pp. 415–426 in Fish Culture in Fisheries Management, R.H. Stroud, ed. Bethesda, MD: American Fisheries Society.

Mueller-Solger, A. 2010. Drivers of change in the San Francisco Estuary. Presentation to the NRC Committee/Panel. December 8, 2010. San Francisco, CA.

Murphy, D.D., and P.S. Weiland. 2011. The route to best science implementation of the Endangered Species Act's consultation mandate: The benefits of structured effects analysis. Environmental Management 47(2):161–172.

NOAA (National Oceanic and Atmospheric Administration). 2008. Federal Columbia River Power System Biological Opinion. Available online at: *http://www.fws.gov/pacific/finalbiop/BiOp.pdf.* Last accessed March 10, 2011.

NOAA. 2010. Endangered Species Act Section 7(a)(2) Consultation Supplemental Biological Opinion: Supplemental Consultation on Remand for Operation of the Federal Columbia River Power System, 11 Bureau of Reclamation Projects in the Columbia Basin and ESA Section 10(a)(I)(A) Permit for Juvenile Fish Transportation Program. Available online at: *https://pcts.nmfs.noaa.gov/pls/pctspub/sxn7.pcts_upload.download?p_file=*

F25013/201002096_FCRPS Supplemental_2010_05-20.pdf. Last accessed April 25, 2011.

NRC (National Research Council). 1995. Science and the Endangered Species Act. Washington, D.C.: National Academy Press.

NRC. 2000. Ecological Indicators for the Nation. Washington, D.C.: National Academy Press.

NRC. 2002. Ecological Dynamics on Yellowstone's Northern Range. Washington, D.C.: National Academy Press.

NRC. 2003. Adaptive Monitoring and Assessment for the Comprehensive Everglades Restoration Plan. Washington, D.C.: The National Academies Press.

NRC. 2004a. Endangered and Threatened Fishes in the Klamath River Basin: causes of Decline and Strategies for Recovery. Washington, D.C.: The National Academies Press.

NRC. 2004b. Atlantic Salmon in Maine. Washington, D.C.: The National Academies Press.

NRC. 2005. Developing a Research and Restoration Plan for Arctic-Yukon Kuskokwim (Western Alaska) Salmon. Washington, D.C.: The National Academies Press.

NRC. 2006. Progress Toward Restoring the Everglades: The First Biennial Review. Washington, D.C.: The National Academies Press.

NRC. 2008. Progress Toward Restoring the Everglades: The Second Biennial Review. Washington, D.C.: The National Academies Press.

NRC. 2009. Science and Decisions. Washington, D.C.: The National Academies Press.

NRC. 2010a. Adapting to the Impacts of Climate Change. America's Climate Choices: Panel on Adapting to the Impacts of Climate Change. Washington, D.C.: The National Academies. Press.

NRC. 2010b. A Scientific Assessment of Alternatives for Reducing Water Management Effects on Threatened and Endangered Fishes in California's Bay-Delta. Washington, D.C.: The National Academies Press.

NRC. 2010c. Progress Toward Restoring the Everglades: The Third Biennial Review, 2010. Washington, D.C.: The National Academies Press.

National Research Defense Council. Map of the Bay Delta Watershed. Available online at: *http://www.nrdc.org/greengate/water/diverted.html.* Last accessed on April 11, 2011.

Orians, G., and D. Policansky. 2009. Scientific bases of macroenvironmental indicators. Annual Reviews of Environment and Resources 34:375–404.

Pine, W.E., S.J.D. Martell, C. J. Walters, J.E. Kitchell. 2009. Counterintuitive responses of fish populations to managements actions: some common causes and implications for predictions based on ecosystem modeling. Fisheries 34(4):165–180.

RECOVER. 2004. CERP Monitoring and Assessment Plan: Part 1, Monitoring and Supporting Research. Restoration Coordination and Verification, c/o U.S. Army Corps of Engineers, Jacksonville District, Jacksonville, FL and

South Florida Water Management District, West Palm Beach, FL. Available online at: *http://www.evergladesplan.org/pm/recover/recover_map.aspx.* Last accessed March 10, 2011.

RECOVER. 2006a. Comprehensive Everglades Restoration Plan Adaptive Management Strategy. Restoration Coordination and Verification Program, c/o U.S. Army Corps of Engineers, Jacksonville District, Jacksonville, FL and South Florida Water Management District, West Palm Beach, FL. April 2006. Available online at: *http://www.evergladesplan.org/pm/recover/recover_docs/am/rec_am_stategy_brochure.pdf.* Last accessed March 10, 2011.

RECOVER. 2006b. CERP Monitoring and Assessment Plan (MAP): Part 2, 2006 Assessment Strategy for the MAP. Restoration Coordination and Verification, c/o U.S. Army Corps of Engineers, Jacksonville District, Jacksonville, FL and South Florida Water Management District, West Palm Beach, FL.

RECOVER. 2009. CERP Monitoring and Assessment Plan (MAP). Restoration Coordination and Verification, C/O U.S. Army Corps of Engineers, Jacksonville District, Jacksonville, FL and South Florida Water Management District, West Palm Beach, FL. Available online at: *http://www.evergladesplan.org/pm/recover/recover_map.aspx.* Last accessed March 10, 2011.

RECOVER. 2010. CERP Adaptive Management Integration Guide (draft), (2010). C/O U.S. Army Corps of Engineers, Jacksonville District, Jacksonville, FL and South Florida Water Management District, West Palm Beach, FL. Available online at: *http://www.evergladesplan.org/pm/pm_docs/adaptive_mgmt/090110_cerp_amig_v3_4.pdf.* Last accessed March 10, 2011.

Redmond, K.T., and R.W. Koch. 1991. Surface climate and streamflow variability in the western United States and their relationship to large-scale circulation indices. Resources Research 27(9):2381–2399.

Richardson, Jr., J.J. 2010. Conservation easements and adaptive management. Sea Grant Law and Policy Journal, 3(1):31–58.

Rieman B., J.T. Peterson, J. Clayton, P. Howell, R. Thurow, W. Thompson, and D. Lee. 2001. Evaluation of potential effects of federal land management alternatives on trends of salmonids and their habitats in the interior Columbia River basin. Forest Ecology and Management 153:43–62.

Runge, M.C., E. Bean, D.R. Smith, and S. Kokos. 2011. Non-native fish control below Glen Canyon Dam—Report from a structured decision-making project: U.S. Geological Survey Open-File Report 2011–1012. Available online at: *http://pubs.usgs.gov/of/2011/1012/.* Last accessed March 10, 2011.

Sagouspe, J. 2010. Letter to the Honarable David Hayes from Jean Sagouspe, President, Westlands Water District, to David Hayes, Deputy Secretary of the Interior, November 22nd, 2010. Available online at: *http://www.westlandswater.org/wwd%5Cpr%5C2010-11-22_sagouspe_ltr_to_hayes.pdf.* Last accessed on June 21, 2011.

Schemel, L.E., T.R. Sommer, A.B. Muller-Solger, and W.C. Harrell. 2004.

Hydrologic variability, water chemistry, and phytoplankton biomass in a large floodplain of the Sacramento River, CA. Hydrobiologia 513:129–139.

Scholz, J.T., and B. Stiftel. 2005. The challenges of adaptive governance. Pp. 1–11 in Adaptive Governance and Water Conflict: New Institutions for Collaborative Planning. Washington, D.C.: Resources for the Future.

Sommer, T., C. Armor, R. Baxter, R. Breuer, L. Brown, M. Chotkowski, S. Culberson, F. Feyrer, M. Gingras, B. Herbold, W. Kimmerer, A. Mueller-Solger, M. Nobriga, and K. Souza. 2007. The collapse of pelagic fishes in the upper San Francisco Estuary. Fisheries 32(6):270–277.

Suddeth, R., J.F. Mount, and J.R. Lund. 2010. Levee decisions and sustainability for the Sacramento-San Joaquin Delta. San Francisco Estuary and Watershed Science 8(2):1–23.

Susskind, L., A.E. Camacho, and T. Schenk. 2010. Collaborative planning and adaptive management in Glen Canyon: a cautionary tale. Columbia Journal of Environmental Law: 35:1–54.

Tanaka, S. K., C. R. Connell, K. Madani, J. Lund, E. Hanak, and J. Medellin-Azuara 2008. The economic costs and adaptations for alternative Delta regulations. Chapter in Comparing Futures for the Sacramento-San Joaquin Delta, J. Lund, E. Hanak, W. Fleenor, W. Bennett, R. Howitt, J. Mount, and P. Moyle, ed. San Francisco, CA: Public Policy Institute of California.

Thompson, J. 1957. Settlement geography of the Sacramento-San Joaquin Delta, California. Ph.D. Dissertation. Available from: Stanford University.

USACE (U.S. Army Corps of Engineers) and SFWMD (South Florida Water Management District). 1999. The Central and Southern Florida Flood Control Project Comprehensive Review Study Final Integrated Feasibility Report and Programmatic Impact Statement (PEIS). West Palm Beach, FL: South Florida Water Management District. Available online at: *http://www.evergladesplan.org/pub/restudy_eis.cfm#mainreport.* Last accessed March 10, 2011.

USBR (U.S. Bureau of Reclamation). 2008. Appendix R. Sensitivity of Future Central Valley Project and State Water Project Operations to Potential Climate Change and Associated Sea Level Rise, OCAP-BA, July, 2008. Available online at: *http://www.usbr.gov/mp/cvo/OCAP/sep08_docs/Appendix_R.pdf.* Last accessed March 10, 2011.

Uusitalo, L. 2007. Advantages and challenges of Bayesian networks in environmental modelling. Ecological Modelling 203:312–318.

Wagner, R.W., M. Stacey, L.R. Brown, and M. Dettinger. 2011. Statistical models of temperature in the Sacramento–San Joaquin Delta under climate-change scenarios and ecological implications. Estuaries and Coasts DOI 10.1007/s12237-010-9369-z. Available on line at: *http://tenaya.ucsd.edu/ ~dettinge/wagner_delta_temps.pdf.* Last accessed April 13, 2011.

Walters, C.J. 2007. Is adaptive management helping to solve fisheries problems? AMBIO: A Journal of the Human Environment 36(4):304–307.

Williams, B.K., R.C. Szaro, and C.D. Shapiro. 2009. Adaptive Management:

The U.S. Department of the Interior Technical Guide. Washington, D.C.: Department of the Interior, Adaptive Management Working Group.

Appendixes

Appendix A
Statement of Task

At the request of the U.S. Departments of Interior and Commerce, a National Research Council panel of independent experts will review the draft Bay Delta Conservation Plan (BDCP), which is being prepared through a collaboration of state, federal, and local water agencies, state and federal fish agencies, environmental organizations, and other interested parties to restore the California Bay-Delta ecosystem and protect water supplies, i.e., provide for both species/habitat protection and improved reliability of water supplies.

Specifically, the panel will provide a short report assessing the adequacy of the use of science and adaptive management in the initial public draft of the Bay Delta Conservation Plan (BDCP) by April 2011. This draft, which is currently scheduled for release on November 24th, 2010, will identify a set of water flow and habitat restoration actions to contribute to the recovery of endangered and sensitive species and their habitats in California's Sacramento-San Joaquin Delta while improving water supply reliability.

The panel's review will be related to but be conducted separately from the on-going, more broadly focused National Research Council *Committee on Sustainable Water and Environmental Management in the California Bay-Delta.* The panel's report is expected to contribute to the broader study which will be completed in late 2011.

Appendix B
BDCP Steering Committee Members and Planning Agreement Signature Dates

Entities	Original Signature Date	Amendment Signature Date
State and Federal Agencies		
California Natural Resources Agency	October 24, 2006	October 27, 2009
California Department of Water Resources	November 14, 2006	December 3, 2009
State Water Resources Control Board (*ex officio*)	See Note	See Note
U.S. Bureau of Reclamation	November 13, 2006	October 30, 2009
U.S. Army Corps of Engineers (*ex officio*)	See Note	See Note
Potential Regulated Entities (PREs)		
Kern County Water Agency	December 6, 2006	January 29, 2010
Metropolitan Water District of Southern California	November 2, 2006	December 3, 2009
Mirant Delta, LLC	December 6, 2006	October 5, 2009
San Luis & Delta-Mendota Water Authority	December 6, 2006	December 6, 2009
Santa Clara Valley Water District	November 20, 2006	November 30, 2009
Westlands Water District	December 6, 2006	December 1, 2009
Zone 7 Water Agency	October 26, 2006	November 30, 2009
Environmental Organizations		
American Rivers	November 8, 2006	January 21, 2010
Defenders of Wildlife	March 15, 2007	January 29, 2010
Environmental Defense Fund	October 30, 2006	January 21, 2010
Natural Heritage Institute	October 25, 2006	November 3, 2009
The Nature Conservancy	November 14, 2006	December 1, 2009
The Bay Institute	July 26, 2007	December 7, 2009
Other Member Agencies		
California Farm Bureau Federation	March 30, 2007	November 11, 2009
Contra Costa Water District	August 3, 2007	January 4, 2010
Friant Water Authority	March 9, 2009	November 18, 2009
North Delta Water Agency	March 12, 2009	October 5, 2009
Fishery Agencies		
California Department of Fish and Game (*ex officio*)	October 24, 2006	October 5, 2009
U.S. Fish and Wildlife Service (*ex officio*)	November 6, 2006	December 3, 2009
National Marine Fisheries Service (*ex officio*)	November 14, 2006	December 3, 2009
Other *Ex Officio* Member Agencies		
Delta Stewardship Council		
Note: The SWRCB and USACE are not signatories of the Planning Agreement.		

SOURCE: Draft BDCP (November 2010).

Appendix C
BDCP Proposed Covered Species and
Associated Habitats

No.	Common Name/ Scientific Name	Status (Federal/ State/CNPS)[1]	Natural Communities Supporting Species Habitat
Fish (11 species)			
1	Central Valley steelhead *Oncorhynchus mykiss* DPS	T/-/- DPS Critical Habitat, Recovery Plan[11]	Tidal perennial aquatic, tidal mud-flats, tidal brackish emergent wetland, tidal freshwater emergent wetland
2	Sacramento River winter-run Chinook salmon *Oncorhynchus tshawytscha* Evolutionarily Significant Unit (ESU)	E/E/- ESU Critical Habitat, Recovery Plan[11, 12]	Tidal perennial aquatic, tidal mud-flats, tidal brackish emergent wetland, tidal freshwater emergent wetland
3	Central Valley spring-run Chinook salmon *Oncorhynchus tshawytscha* ESU	T/T/- ESU Critical Habitat, Recovery Plan[11, 13]	Tidal perennial aquatic, tidal mud-flats, tidal brackish emergent wetland, tidal freshwater emergent wetland
4	Central Valley fall- and late fall-run Chinook salmon *Oncorhynchus tshawytscha*	-/SSC/- Recovery Plan[13]	Tidal perennial aquatic, tidal mud-flats, tidal brackish emergent wetland, tidal freshwater emergent wetland
5	Delta smelt *Hypomesus transpacificus*	T/T/- Critical Habitat, Recovery Plan[13]	Tidal perennial aquatic, tidal mud-flats, tidal brackish emergent wetland, tidal freshwater emergent wetland
6	Longfin smelt *Spirinchus thaleichthys*	-/T/- Recovery Plan[13]	Tidal perennial aquatic, tidal mud-flats, tidal brackish emergent wetland, tidal freshwater emergent wetland
7	Sacramento splittail *Pogonichthys macrolepidotus*	-/SSC/- Recovery Plan[13]	Tidal perennial aquatic, tidal mud-flats, tidal brackish emergent wetland, tidal freshwater emergent wetland
8	White sturgeon *Acipenser transmontanus*	-/-/-	Tidal perennial aquatic, tidal mud-flats, tidal brackish emergent wetland, tidal freshwater emergent wetland

No.	Common Name/ Scientific Name	Status (Federal/ State/CNPS)[1]	Natural Communities Supporting Species Habitat
9	North American green sturgeon *Acipenser medirostris* Southern DPS	T/SSC/- Southern DPS *Proposed* Critical Habitat, Recovery Plan[13]	Tidal perennial aquatic, tidal mudflats, tidal brackish emergent wetland, tidal freshwater emergent wetland
10	Pacific lamprey *Entosphenus tridentatus*	-/-/-	Tidal perennial aquatic, tidal mudflats, tidal brackish emergent wetland, tidal freshwater emergent wetland
11	River lamprey *Lampetra ayresii*	-/-/-	Tidal perennial aquatic, tidal mudflats, tidal brackish emergent wetland, tidal freshwater emergent wetland
Mammals (6 species)			
12	San Joaquin kit fox *Vulpes macrotis mutica*	E/T/-Recovery Plan[2]	Grassland, Agricultural habitats
13	Riparian woodrat *Neotoma fuscipes riparia*	E/SSC/- Recovery Plan[2]	Valley/foothill riparian
14	Salt marsh harvest mouse *Reithrodontomys raviventris*	E/E,FP/- Recovery Plan[3, 4]	Tidal brackish emergent wetland, managed wetlands, grassland
15	Riparian brush rabbit *Sylvilagus bachmani riparius*	E/E/-Recovery Plan[2]	Valley/foothill riparian
16	Townsend's big-eared bat *Corynorhinus townsendii*	-/SSC/-	All natural communities
17	Suisun shrew *Sorex ornatus sinuosus*	-/SSC/- Recovery Plan[3]	Tidal brackish emergent wetland, managed wetlands
Birds (12 species)			
18	Tricolored blackbird *Agelaius tricolor*	-/SSC/-	Tidal brackish emergent wetland, tidal freshwater emergent wetland, valley/foothill riparian, alkali seasonal wetland complex, managed wetlands, other natural seasonal wetlands, grassland, agricultural habitats
19	Suisun song sparrow *Melospiza melodia maxillaries*	-/SSC/- Recovery Plan[4]	Tidal brackish emergent wetland, tidal freshwater emergent wetland, managed wetlands
20	Yellow-breasted chat *Icteria virens*	-/SSC/-	Valley/foothill riparian
21	Least Bell's vireo *Vireo bellii pusillus*	E/E/- Recovery Plan[5]	Valley/foothill riparian

No.	Common Name/ Scientific Name	Status (Federal/ State/CNPS)[1]	Natural Communities Supporting Species Habitat
22	Western burrowing owl *Athene cunicularia hypugaea*	-/SSC/-	Grassland, alkali seasonal wetland complex, vernal pool complex, managed wetland, other natural seasonal wetlands, agricultural habitats
23	Western yellow-billed cuckoo *Coccyzus americanus occidentalis*	C/E/-	Valley/foothill riparian
24	California least tern *Sternula antillarum browni*	E/E/-Recovery Plan[6]	Tidal perennial aquatic
25	Greater sandhill crane *Grus canadensis tabida*	-/T,FP/-	Agricultural habitats, alkali seasonal wetland complex, vernal pool complex, managed wetlands, other natural seasonal wetlands, grassland
26	California black rail *Laterallus jamaicensis coturniculus*	-/T,FP/- Recovery Plan[4]	Tidal brackish emergent wetland, tidal freshwater emergent wetland, nontidal freshwater permanent emergent wetland
27	California clapper rail *Rallus longirostris obsoletus*	E/E,FP/- Recovery Plan[3, 4]	Tidal brackish emergent wetland
28	Swainson's hawk *Buteo swainsoni*	-/T/-	Valley/foothill riparian, agricultural habitats, grassland, alkali seasonal wetland complex, vernal pool complex, managed wetlands, other natural seasonal wetlands
29	White-tailed kite *Elanus leucurus*	-/FP/-	Valley/foothill riparian, agricultural habitats, grassland, alkali seasonal wetland complex, vernal pool complex, managed wetlands, other natural seasonal wetlands
Reptiles (2 species)			
30	Giant garter snake *Thamnophis gigas*	T/T/-Recovery Plan[6]	Tidal perennial aquatic, tidal freshwater emergent wetland, nontidal perennial aquatic, nontidal freshwater permanent emergent wetland, alkali seasonal wetland complex, vernal pool complex, managed wetlands, other natural seasonal wetlands, grassland, agricultural habitats

No.	Common Name/ Scientific Name	Status (Federal/ State/CNPS)[1]	Natural Communities Supporting Species Habitat
31	Western pond turtle *Actinemys* (formerly *Clemmys* and *Emys*) *marmorata*	-/SSC/-	Tidal perennial aquatic, tidal freshwater emergent wetland, tidal brackish emergent wetland, nontidal perennial aquatic, nontidal freshwater permanent emergent wetland, valley/foothill riparian, alkali seasonal wetland complex, vernal pool complex, managed wetlands, other natural seasonal wetlands, grassland, agricultural habitats
Amphibians (3 species)			
32	California red-legged frog *Rana draytonii*	T/SSC/-Critical Habitat, Recovery Plan[8]	Valley/foothill riparian, nontidal freshwater permanent emergent wetland, tidal freshwater emergent wetland, nontidal perennial aquatic, managed wetlands, grassland, alkali seasonal wetland complex, vernal pool complex, other natural seasonal wetlands, agricultural habitats
33	Western spadefoot toad *Spea hammondii*	-/SSC/- Recovery Plan[9]	Grassland, alkali seasonal wetland complex, vernal pool complex, other natural seasonal wetlands, nontidal perennial aquatic
34	California tiger salamander *Ambystoma californiense* Central Valley Distinct Population Segment (DPS)	T/T/-Central Valley DPS Critical Habitat	Vernal pool complex, alkali seasonal wetland complex, other natural seasonal wetlands, grassland
Invertebrates (8 species)			
35	Lange's metalmark butterfly *Apodemia mormo langei*	E/-/-Recovery Plan[15]	Inland dune scrub
36	Valley elderberry longhorn beetle *Desmocerus californicus dimorphus*	T/-/-Recovery Plan[14]	Valley/foothill riparian, grassland
37	Vernal pool tadpole shrimp *Lepidurus packardi*	E/-/-Critical Habitat Recovery Plan[9]	Vernal pool complex
38	Conservancy fairy shrimp *Branchinecta conservatio*	E/-/-Critical Habitat Recovery Plan[9]	Vernal pool complex
39	Longhorn fairy shrimp *Branchinecta longiantenna*	E/-/-Recovery Plan[9]	Vernal pool complex
40	Vernal pool fairy shrimp *Branchinecta lynchi*	T/-/-Critical Habitat Recovery Plan[9]	Vernal pool complex

No.	Common Name/ Scientific Name	Status (Federal/ State/CNPS)[1]	Natural Communities Supporting Species Habitat
41	Midvalley fairy shrimp *Branchinecta mesovallensis*	-/-/- Recovery Plan[9]	Vernal pool complex
42	California linderiella *Linderiella occidentalis*	-/-/- Recovery Plan[9]	Vernal pool complex
Plants (21 species)			
43	Alkali milk-vetch *Astragalus tener* var. *tener*	-/-/1B Recovery Plan[9]	Vernal pool complex
44	Heartscale *Atriplex cordulata*	-/-/1B	Alkali seasonal wetland complex, vernal pool complex, grassland
45	Brittlescale *Atriplex depressa*	-/-/1B	Alkali seasonal wetland complex, vernal pool complex, grassland
46	San Joaquin spearscale *Atriplex joaquiniana*	-/-/1B	Alkali seasonal wetland complex, vernal pool complex, grassland
47	Slough thistle *Cirsium crassicaule*	-/-/1B	Valley/foothill riparian
48	Suisun thistle *Cirsium hydrophilum var.hydrophilum*	E/-/1B Critical Habitat Recovery Plan[4]	Tidal brackish emergent wetland
49	Soft bird's-beak *Cordylanthus mollis ssp. mollis*	E/R/IB Critical Habitat Recovery Plan[4]	Tidal brackish emergent wetland
50	Dwarf downingia *Downingia pusilla*	-/-/2	Vernal pool complex
51	Delta button-celery *Eryngium racemosum*	-/E/1B	Alkali seasonal wetland complex, vernal pool complex, valley/foothill riparian, grassland
52	Contra Costa wallflower *Erysimum capitatum var. angustatum*	E/E/1B Critical Habitat Recovery Plan[15]	Inland dune scrub
53	Boggs Lake hedge-hyssop *Gratiola heterosepala*	-/E/1B Recovery Plan[9]	Vernal pool complex
54	Carquinez goldenbush *Isocoma arguta*	-/-/1B	Alkali seasonal wetland complex, grassland
55	Delta tule pea *Lathyrus jepsonii var. jepsonii*	-/-/1B Recovery Plan[4]	Tidal brackish emergent wetland, tidal freshwater emergent wetland, valley/foothill riparian
56	Legenere *Legenere limosa*	-/-/1B Recovery Plan[9]	Vernal pool complex
57	Heckard's peppergrass *Lepidium latipes var. heckardii*	-/-/1B	Vernal pool complex
58	Mason's lilaeopsis *Lilaeopsis masonii*	-/R/1B	Tidal mudflats, tidal brackish emergent wetland, tidal freshwater emergent wetland, valley/foothill riparian

No.	Common Name/ Scientific Name	Status (Federal/ State/CNPS)[1]	Natural Communities Supporting Species Habitat
59	Delta mudwort Limosella subulata	-/-/2	Tidal mudflats, tidal brackish emergent wetland, tidal freshwater emergent wetland, valley/foothill riparian
60	Antioch Dunes evening-primrose Oenothera deltoides ssp. howellii	E/E/1B Critical Habitat Recovery Plan[15]	Inland dune scrub
61	Side-flowering skullcap Scutellaria lateriflora	-/-/2	Valley/foothill riparian
62	Suisun Marsh aster Symphyotrichum (formerly Aster lentus) lentum	-/-/1B	Tidal brackish emergent wetland, tidal freshwater emergent wetland, valley/foothill riparian
63	Caper-fruited tropidocarpum Tropidocarpum capparideum	-/-/1B	Grassland

Note: This table provides the current list of proposed covered species. Additional species may be added and some of the species presented here may be removed from the covered species list as per continuing development of the BDCP.

1Status:

Federal

E = Listed as endangered under ESA

T = Listed as threatened under ESA

C = Candidate for listing under ESA

State

E = Listed as endangered under CESA

T = Listed as threatened under CESA

R = Listed as rare under the California Native Plant Protection Act

SSC = California species of special concern

FP = Fully protected under the California Fish and Game Code

California Native Plant Society (CNPS)

1B = rare or endangered in California and elsewhere

2 = rare and endangered in California, more common elsewhere

2U.S. Fish and Wildlife Service. 1998. Recovery plan for upland species of the San Joaquin Valley, California. Region 1, Portland,

OR. 319 pp.

3U.S. Fish and Wildlife Service. 1984. Salt marsh harvest mouse and California clapper rail recovery plan. Portland, OR.

4U.S. Fish and Wildlife Service. 2009. Draft Recovery Plan for Tidal Marsh Ecosystems of Northern and Central California.

Sacramento, California. xviii+636 pp.

5U.S. Fish and Wildlife Service. 1998. Draft recovery plan for the least Bell's vireo. U.S. Fish and Wildlife Service, Portland, OR.

139 pp.

6U.S. Fish and Wildlife Service. 1985. Recovery plan for the California least tern, Sterna antillarum browni. U.S. Fish and

Wildlife Service, Portland, OR. 112 pp.

7U.S. Fish and Wildlife Service. 1999. Draft Recovery Plan for the Giant Garter Snake (Thamnopsis gigas). U.S. Fish and Wildlife

Service, Portland, Pregon. ix+192 pp.

8U.S. Fish and Wildlife Service. 2002. Recovery Plan for the California Red-legged Frog (Rana aurora draytonii). U.S. fish and

Wildlife Service, Portland, Oregon. viii+173 pp.

9U.S. Fish and Wildlife Service. 2005. Recovery Plan for Vernal Pool Ecosystems of California and Southern Oregon. Portland,
Oregon. xxvi + 606 pages.
10California Tiger Salamander distinct population segments are federally listed as endangered in Sonoma and Santa Barbara
counties.
11National Marine Fisheries Service. 2009. Public Draft Recovery Plan for the Evolutionarily Significant Units of Sacramento
River Winter-run Chinook Salmon and Central Valley Spring-run Chinook Salmon and the Distinct Population Segment of
Central Valley Steelhead. Sacramento Protected Resources Division. October 2009.
12National Marine Fisheries Service. 1997. NMFS Proposed Recovery Plan for the Sacramento River winter-run Chinook Salmon.
NMFS Southwest Region. Long Beach, CA.
13U.S. Fish and Wildlife Service. 1995. Sacramento-San Joaquin Delta Native Fishes Recovery Plan. U.S. Fish and Wildlife
Service, Portland, Oregon.
14U.S. Fish and Wildlife Service. 1984. Valley elderberry longhorn beetle Recovery Plan. U.S. Fish and Wildlife Service, Portland,
Oregon. 62 pp.
15U.S. Fish and Wildlife Service. 1984. Revised recovery plan for three endangered species endemic to Antioch Dunes, California.
16U.S. Fish and Wildlife Service, Portland, Oregon

SOURCE: BDCP (Bay Delta Conservation Plan Steering Committee). 2010. Bay Delta Conservation Plan Working Draft. November 18. Available online at: *http://www.resources.ca.gov/bdcp/*. Last accessed April 26, 2011.

Appendix D
Possible Causal Connections in Suppression of Populations of Endangered Suckers in Upper Klamath Lake

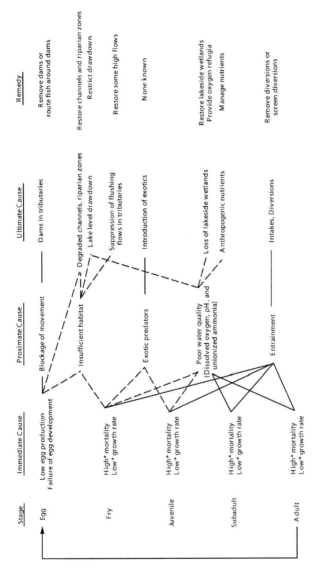

NOTE: Solid line = connection verified scientifically; dashed line = connection under study. "High mortality" and "low growth rate" are relative to rates in stable populations.

SOURCE: NRC, 2004a

Appendix E
BDCP Adaptive Management
Process Framework

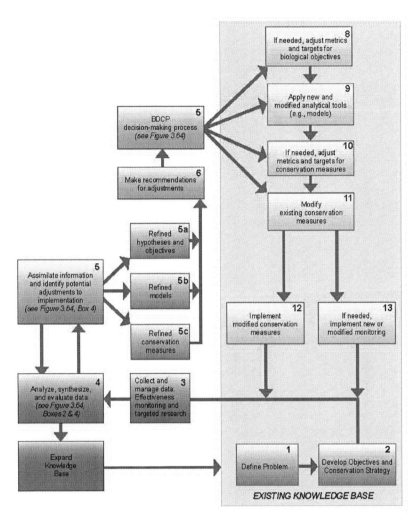

SOURCE: Draft BDCP (November 2010).

Appendix F
Water Science and Technology Board

DONALD I. SIEGEL, Chair, Syracuse University, New York
LISA M. ALVAREZ-COHEN, University of California, Berkeley
EDWARD J. BOUWER, Johns Hopkins University, Baltimore, Maryland
YU-PING CHIN, Ohio State University, Columbus
OTTO C. DOERING III, Purdue University, West Lafayette, Indiana
M. SIOBHAN FENNESSY, Kenyon College, Gambier, Ohio
BEN GRUMBLES, Clean Water America Alliance, Washington, DC
GEORGE R. HALLBERG, The Cadmus Group, Watertown, Massachusetts
KENNETH R. HERD, Southwest Florida Water Management District,
  Brooksville
GEORGE M. HORNBERGER, Vanderbilt University, Nashville, Tennessee
LARRY LARSON, Association of State Floodplain Managers, Madison,
  Wisconsin
DAVID H. MOREAU, University of North Carolina, Chapel Hill
DENNIS D. MURPHY, University of Nevada, Reno
MARYLYNN V. YATES, University of California, Riverside

Staff

STEPHEN D. PARKER, Director
JEFFREY JACOBS, Scholar
LAURA J. EHLERS, Senior Staff Officer
STEPHANIE E. JOHNSON, Senior Staff Officer
LAURA J. HELSABECK, Staff Officer
M. JEANNE AQUILINO, Financial/Administrative Associate
ELLEN A. DE GUZMAN, Research Associate/Senior Program Associate
ANITA A. HALL, Senior Program Associate
MICHAEL STOEVER, Research Associate
SARAH BRENNAN, Senior Project Assistant

Appendix G
Ocean Studies Board

DONALD F. BOESCH, *Chair,* University of Maryland Center for Environmental Science, Cambridge
EDWARD A. BOYLE, Massachusetts Institute of Technology, Cambridge
CORTIS K. COOPER, Chevron Corporation, San Ramon, CA
JORGE E. CORREDOR, University of Puerto Rico, Mayaguez
KEITH R. CRIDDLE, University of Alaska Fairbanks, Juneau
JODY W. DEMING, University of Washington, Seattle
ROBERT HALLBERG, NOAA/GFDL and Princeton University, NJ
DEBRA HERNANDEZ, SECOORA, Mt. Pleasant, SC
ROBERT A. HOLMAN, Oregon State University, Corvallis
KIHO KIM, American University, Washington, DC
BARBARA A. KNUTH, Cornell University, Ithaca, NY
ROBERT A. LAWSON, Science Applications International Corporation, San Diego, CA
GEORGE I. MATSUMOTO, Monterey Bay Aquarium Research Institute, Moss Landing, CA
JAY S. PEARLMAN, The Boeing Company (retired), Port Angeles, WA
ANDREW A. ROSENBERG, Conservation International, Arlington, VA
DANIEL L. RUDNICK, Scripps Institution of Oceanography, La Jolla, CA
ANNE M. TREHU, Oregon State University, Corvallis
PETER L. TYACK, Woods Hole Oceanographic Institution, MA
DON WALSH, International Maritime Incorporated, Myrtle Point, OR
DAWN J. WRIGHT, Oregon State University, Corvallis
JAMES A. YODER, Woods Hole Oceanographic Institution, MA
Ex-Officio
MARY (MISSY) H. FEELEY, ExxonMobil Exploration Company, Houston, TX

Staff
SUSAN ROBERTS, Board Director
CLAUDIA MENGELT, Senior Program Officer
KIM WADDELL, Senior Program Officer
DEBORAH GLICKSON, Senior Program Officer
MARTHA MCCONNELL, Program Officer
SHUBHA BANSKOTA, Financial Associate
PAMELA LEWIS, Administrative Coordinator
SHERRIE FORREST, Research Associate
JEREMY JUSTICE, Senior Program Assistant
EMILY OLIVER, Program Assistant
PETER THOMPSON, Mirzayan Fellow

Appendix H
Panel Biographical Information

HENRY J. VAUX, JR., *Chair*, is Professor Emeritus of Resource Economics at both the University of California in Berkley and Riverside. He is also Associate Vice President Emeritus of the University of California system. He also previously served as director of California's Center for Water Resources. His principal research interests are the economics of water use, water quality, and water marketing. Prior to joining the University of California, he worked at the Office of Management and Budget and served on the staff of the National Water Commission. Dr. Vaux has served on the NRC committees on Assessment of Water Resources Research, Western Water Management, and Ground Water Recharge, and Sustainable Underground Storage of Recoverable Water. He was chair of the Water Science and Technology Board from 1994 to 2001. He is a National Associate of The National Academies. Dr. Vaux received an A.B. from the University of California, Davis in biological sciences, an M.A. in natural resource administration, and an M.S. and Ph.D. in economics from the University of Michigan.

MICHAEL E. CAMPANA is Professor of Geosciences at Oregon State University (OSU), former Director of its Institute for Water and Watersheds, and Emeritus Professor of Earth and Planetary Sciences at the University of New Mexico. Prior to joining OSU in 2006 he held the Albert J. and Mary Jane Black Chair of Hydrogeology and directed the Water Resources Program at the University of New Mexico, was a research hydrologist at the Desert Research Institute, and taught in the University of Nevada-Reno's Hydrologic Sciences Program. He has supervised 70 graduate students. His research and interests include hydrophilanthropy, water resources management and policy, communications, transboundary water resources, hydrogeology, and environmental fluid mechanics, and he has published on a variety of topics. Dr. Campana was a Fulbright Scholar to Belize and a Visiting Scientist at Research Institute for Groundwater (Egypt) and the IAEA in Vienna. Central America and the South Caucasus are the current foci of his international work. He has served on six NRC-NAS committees. Dr. Campana is Founder, President, and Treasurer of the Ann Campana Judge Foundation (www.acjfoundation.org), a 501(c)(3) charitable foundation that funds and undertakes projects related to water, sanitation, and hygiene (WASH) in Central America. He operates the WaterWired blog and Twitter. He earned a B.S. in geology from the College of William and Mary, and M.S. and Ph.D. degrees in hydrology from the University of Arizona.

JEROME B. GILBERT is a consulting engineer and founder of J. Gilbert, Inc. His interests include integrated water supply and water quality planning and management. Mr. Gilbert has managed local and regional utilities, and he has developed basin/watershed water quality and protection plans. He has supervised California's water rights and water quality planning and regulatory activities, chaired the San Francisco Bay Regional Water Quality Control Board, and led national and international water and water research associations. Areas of experience include: authorship of state and national water legislation on water rights, pollution control, water conservation and urban water management; optimization of regional water project development; groundwater remediation and conjunctive use; economic analysis of alternative water improvement projects; and planning of multipurpose water management efforts including remediation. He has served on national panels related to control and remediation of ground and surface water contamination, and the National Drinking Water Advisory Council. Mr. Gilbert is a member of the National Academy of Engineering. He received his B.S. from the University of Cincinnati and an M.S. from Stanford University.

ALBERT E. GIORGI is President and Senior Fisheries Scientist at BioAnalysts, Inc. in Redmond, Washington. He has been conducting research on Pacific Northwest salmonid resources since 1982. Prior to 1982, he was a research scientist with NOAA in Seattle, Washington. He specializes in fish passage migratory behavior, juvenile salmon survival studies, and biological effects of hydroelectric facilities and operation. His research includes the use of radio telemetry, acoustic tags, and PIT-tag technologies. In addition to his research, he acts as a technical analyst and advisor to public agencies and private parties. He regularly teams with structural and hydraulic engineers in the design and evaluation of fishways and fish bypass systems. He also has served on the NRC Committee on Water Resources Management, Instream Flows, and Salmon Survival in the Columbia River. He received his B.A. and M.A. in biology from Humboldt State University and his Ph.D. in fisheries from the University of Washington.

ROBERT J. HUGGETT is an independent consultant and Professor Emeritus and former Chair of the Department of Environmental Sciences, Virginia Institute of Marine Sciences at the College of William and Mary, where he was on the faculty for more than 20 years. He also served as Professor of Zoology and Vice President for Research and Graduate Studies at Michigan State University from 1997 to 2004. Dr. Huggett is an expert in aquatic biogeochemistry and ecosystem management whose research involved the fate and effects of hazardous substances in aquatic systems. From 1994 to 1997, he was the Assistant Administrator for Research and Development for the U.S. Environmental Protection Agency (EPA, where his responsibilities included planning and directing the agency's research program. During his time at the EPA, he served as Vice Chair of the Committee on Environment and Natural Resources and Chair of the Subcommittee on Toxic Substances and Solid Wastes, both of the White House Office of Science and Technology Policy. Dr. Huggett founded the EPA Star Competitive Research Grants program and the EPA Star Graduate Fellowship program. He has served on the National Research Council's (NRC) Board on Environmental Studies and Toxicology, the Water Science and Technology Board, and numerous study committees on wide ranging topics. Dr. Huggett earned an M.S. in marine chemistry from the Scripps Institution of Oceanography at the University of California at San Diego and completed his Ph.D. in marine science at the College of William and Mary.

CHRISTINE A. KLEIN is the Chesterfield Smith Professor of Law at the University of Florida Levin College of Law, where she has been teaching since 2003. She offers courses on natural resources law, environmental law, water law, and property. Previously, she was a member of the faculty of Michigan State University College of Law, where she served as Environmental Law Program Director. From 1989 to 1993, she was an Assistant Attorney General in the Office of the Colorado Attorney General, Natural Resources Section, where she specialized in water rights litigation. She has published widely on a variety of water law and natural resources law topics. She holds a B.A. from Middlebury College, Vermont; a J.D. from the University of Colorado School of Law; and an LL.M. from Columbia University School of Law, New York.

SAMUEL N. LUOMA is an emeritus Senior Research Hydrologist in the Water Resources Division of the U.S. Geological Survey, where he worked for 34 years. Dr. Luoma's research centers on sediment processes, both natural and human-induced, particularly in the San Francisco Bay area. He served as the first lead on the CALFED Bay-delta program and is the Editor-in-Chief of San Francisco Estuary & Watershed Science. Since 1992, he has published extensively on the bioavailability and ecological effects of metals in aquatic environments. He has helped refine approaches to determine the toxicity of marine and estuarine sediments. In 1999, he was invited to discuss how chemical speciation influences metal bioavailability in sediments for the European Science Foundation. He has served multiple times on the EPA's Science Advisory Board Subcommittee on Sediment Quality Criteria and on several NRC committees. Dr. Luoma received his B.S. and M.S. in zoology from Montana State University, Bozeman, and his Ph.D. in marine biology from the University of Hawaii, Honolulu.

THOMAS MILLER is Professor of Fisheries and Bioenergetics and Population Dynamics at the Chesapeake Biological Laboratory, University of Maryland Center for Environmental Science (UMCES-CBL), where he has been teaching since 1994. Prior to UMCES-CBL, he was a postdoctoral fellow at McGill University, Montreal, Canada, and research specialist with the Center for Great Lakes Studies, University of Wisconsin, Milwaukee. His research focuses on population dynamics of aquatic animals, particularly in understanding recruitment, feeding and bio-physical interactions, and early life history of fish and crustaceans. He has been involved in the development of a Chesapeake Bay fishery ecosystem plan, which includes detailed background information on fisheries, foodwebs, habitats and monitoring required to develop multispecies stock assessments. Most recently, he has developed an interest in the sub-lethal effects of contamination on Chesapeake Bay living resources using population dynamic approaches. He received his B.Sc. (hons) in human and environmental biology from the University of York, UK, and his M.S. in ecology and Ph.D. in zoology and oceanography from North Carolina State University.

STEPHEN G. MONISMITH is Professor of Environmental Fluid Mechanics and directs the Environmental Fluid Mechanics Laboratory at Stanford University. Prior to coming to Stanford, he spent 3 years in Perth (Australia) as a research fellow at the University of Western Australia. Dr. Monismith's research in environmental and geophysical fluid dynamics involves the application of fluid mechanics principles to the analysis of flow processes operating in rivers, lakes, estuaries and the oceans. Making use of laboratory experimentation, numerical modelling, and field measurements, his current research includes studies of estuarine hydrodynamics and mixing processes, flows over coral reefs, wind wave-turbulent flow interactions in the upper ocean, turbulence in density stratified fluids, and physical-biological interactions in phytoplankton and benthic systems. He received his B.S., M.S., and Ph.D. from the University of California at Berkeley.

JAYANTHA OBEYSEKERA directs the Hydrologic & Environmental Systems Modeling Department at the South Florida Water Management District, where he is a lead member of a modeling team dealing with development and applications of computer simulation models for Kissimmee River restoration and the restoration of the Everglades Ecosystem. Prior to joining the South Florida Water Management District, he taught courses in hydrology and water resources at Colorado State University, Fort Collins; George Washington University, Washington, DC; and at Florida Atlantic University, Boca Raton, Florida. Dr. Obeysekera has published numerous research articles in refereed journals in the field of water resources. Dr. Obeysekera has more than 20 years of experience practicing water resources engineering with an emphasis on both stochastic and deterministic modeling. He has taught short courses on modeling in the Dominican Republic, Colombia, Spain, Sri Lanka, and the United States. He was a member of the Surface Runoff Committee of the American Geophysical Union and is currently serving as a member of a Federal Task Group on Hydrologic Modeling. He served as member of NRC's Committee on Further Studies of Endangered and Threatened Fishes in the Klamath River. Dr. Obeysekera has a B.S. degree in civil engineering from University of Sri Lanka; M.E. in hydrology from University of Roorkee, India; and Ph.D. in civil engineering with specialization in water resources from Colorado State University.

HANS W. PAERL is Kenan Professor of Marine and Environmental Sciences, at the University of North Carolina Chapel Hill Institute of Marine Sciences, Morehead City. His research includes microbially mediated nutrient cycling and primary production dynamics of aquatic ecosystems, environmental controls of harmful algal blooms, and assessing the causes and consequences of man-made and climatic (storms, floods) nutrient enrichment and hydrologic alterations of inland, estuarine, and coastal waters. His studies have identified the importance and ecological impacts of atmospheric nitrogen deposition as a new nitrogen source supporting estuarine and coastal eutrophication. He is involved in the development and application of microbial and biogeochemical indicators of aquatic ecosystem condition and change in response to human and climatic perturbations. He heads up the Neuse River Estuary Modeling and Monitoring Program, and ferry-based water quality monitoring program, FerryMon, which employs environmental sensors and a various microbial indicators to assess near real-time ecological condition of the Pamlico Sound System, the nation's second largest estuarine complex. In 2003 he was awarded the G. Evelyn Hutchinson Award by the American Society of Limnology and Oceanography for his work in these fields and their application to interdisciplinary research, teaching and management of aquatic ecosystems. He received his PhD from the University of California-Davis.

MAX J. PFEFFER is International Professor of Development Sociology and senior Associate Dean of the College of Agriculture and Life Sciences at Cornell University. His teaching concentrates on environmental sociology and sociological theory. His research spans several areas including farm labor, rural labor markets, international migration, land use, and environmental planning. The empirical work covers a variety of rural and urban communities, including rural/urban fringe areas. Research sites include rural New York and Central America. He has been awarded competitive grants from the National Institutes of Health, the National Science Foundation, the U.S. Environmental Protection Agency, the U.S. Department of Agriculture's National Research Initiative and its Fund for Rural America, and the Social Science Research Council. Dr. Pfeffer has published a wide range of scholarly articles and has written or co-edited four books. He recently published (with John Schelhas) *Saving Forests, Protecting People? Environmental Conservation in Central America.* He also previously served as the Associate Director of both the Cornell University Agricultural Experiment Station and the Cornell University Center for the Environment. He received his Ph.D. degree in sociology from the University of Wisconsin, Madison.

DESIREE D. TULLOS is Assistant Professor in the Department of Biological and Ecological Engineering, Oregon State University, Corvallis. Dr. Tullos also consulted with Blue Land Water Infrastructure and with Barge, Waggoner, Sumner, and Cannon before joining the faculty at Oregon State University. Her research areas include ecohydraulics, river morphology and restoration, bioassessment, and habitat and hydraulic modeling. She has done work on investigations of biological responses to restoration and engineered applications in riverine ecosystems; development and evaluation of targeted and appropriate bioindicators for the assessment of engineered designs in riverine systems; assessing effects of urban and agricultural activities and management practices on aquatic ecosystem stability in developing countries. She received her B.S. in civil engineering from the University of Tennessee, Knoxville, and her MC.E. in civil engineering and Ph.D. in biological engineering from North Carolina State University, Raleigh.

STAFF

LAURA J. HELSABECK is a Staff Officer with the National Research Council's Water Science and Technology Board. Her interests include the use of scientific information to enhance water policy and management decisions pertaining to water quality and quantity. Since joining the National Research Council, she has directed studies for a variety of topics including the Committee on Challenges and Opportunities in the Hydrology Sciences and the Committee on U.S. Geologic Survey's Water Resources Research. Dr. Helsabeck received her B.A. from Clemson University, her M.S. from Vanderbilt University, and Ph.D. from The Ohio State University in Environmental Science. Her dissertation work, Ibuprofen photolysis: Reaction kinetics, chemical mechanism, and byproduct analysis, was awarded the Ellen C. Gonter Environmental Chemistry Award by the American Chemical Society.

DAVID POLICANSKY is a Scholar with the Board on Environmental Studies and Toxicology at the National Research Council, where he directs studies on applied ecology and natural resource management. He chairs the Advisory Council for the University of Alaska's School of Fisheries and Ocean Sciences and was a 2001 Harriman Scholar on the retracing of the 1899 Harriman Alaska Expedition. His research interests include genetics; evolution; and ecology, including the effects of fishing on fish populations; ecological risk assessment; natural resource management; and how science is used in informing policy. He has directed more than 30 projects at the National Research Council on natural resources and ecological risk assessment.